Fluid Mechanics: A Very Short Introduction

VERY SHORT INTRODUCTIONS are for anyone wanting a stimulating and accessible way into a new subject. They are written by experts, and have been translated into more than 45 different languages.

The series began in 1995, and now covers a wide variety of topics in every discipline. The VSI library currently contains over 700 volumes—a Very Short Introduction to everything from Psychology and Philosophy of Science to American History and Relativity—and continues to grow in every subject area.

Very Short Introductions available now:

ABOLITIONISM Richard S. Newman
THE ABRAHAMIC RELIGIONS Charles L. Cohen
ACCOUNTING Christopher Nobes
ADOLESCENCE Peter K. Smith
THEODOR W. ADORNO Andrew Bowie
ADVERTISING Winston Fletcher
AERIAL WARFARE Frank Ledwidge
AESTHETICS Bence Nanay
AFRICAN AMERICAN RELIGION Eddie S. Glaude Jr
AFRICAN HISTORY John Parker and Richard Rathbone
AFRICAN POLITICS Ian Taylor
AFRICAN RELIGIONS Jacob K. Olupona
AGEING Nancy A. Pachana
AGNOSTICISM Robin Le Poidevin
AGRICULTURE Paul Brassley and Richard Soffe
ALEXANDER THE GREAT Hugh Bowden
ALGEBRA Peter M. Higgins
AMERICAN BUSINESS HISTORY Walter A. Friedman
AMERICAN CULTURAL HISTORY Eric Avila
AMERICAN FOREIGN RELATIONS Andrew Preston
AMERICAN HISTORY Paul S. Boyer
AMERICAN IMMIGRATION David A. Gerber
AMERICAN INTELLECTUAL HISTORY Jennifer Ratner-Rosenhagen
AMERICAN LEGAL HISTORY G. Edward White
AMERICAN MILITARY HISTORY Joseph T. Glatthaar
AMERICAN NAVAL HISTORY Craig L. Symonds
AMERICAN POETRY David Caplan
AMERICAN POLITICAL HISTORY Donald Critchlow
AMERICAN POLITICAL PARTIES AND ELECTIONS L. Sandy Maisel
AMERICAN POLITICS Richard M. Valelly
THE AMERICAN PRESIDENCY Charles O. Jones
THE AMERICAN REVOLUTION Robert J. Allison
AMERICAN SLAVERY Heather Andrea Williams
THE AMERICAN SOUTH Charles Reagan Wilson
THE AMERICAN WEST Stephen Aron
AMERICAN WOMEN'S HISTORY Susan Ware
AMPHIBIANS T. S. Kemp
ANAESTHESIA Aidan O'Donnell
ANALYTIC PHILOSOPHY Michael Beaney
ANARCHISM Colin Ward
ANCIENT ASSYRIA Karen Radner
ANCIENT EGYPT Ian Shaw

ANCIENT EGYPTIAN ART AND ARCHITECTURE Christina Riggs
ANCIENT GREECE Paul Cartledge
THE ANCIENT NEAR EAST Amanda H. Podany
ANCIENT PHILOSOPHY Julia Annas
ANCIENT WARFARE Harry Sidebottom
ANGELS David Albert Jones
ANGLICANISM Mark Chapman
THE ANGLO-SAXON AGE John Blair
ANIMAL BEHAVIOUR Tristram D. Wyatt
THE ANIMAL KINGDOM Peter Holland
ANIMAL RIGHTS David DeGrazia
THE ANTARCTIC Klaus Dodds
ANTHROPOCENE Erle C. Ellis
ANTISEMITISM Steven Beller
ANXIETY Daniel Freeman and Jason Freeman
THE APOCRYPHAL GOSPELS Paul Foster
APPLIED MATHEMATICS Alain Goriely
THOMAS AQUINAS Fergus Kerr
ARBITRATION Thomas Schultz and Thomas Grant
ARCHAEOLOGY Paul Bahn
ARCHITECTURE Andrew Ballantyne
THE ARCTIC Klaus Dodds and Jamie Woodward
ARISTOCRACY William Doyle
ARISTOTLE Jonathan Barnes
ART HISTORY Dana Arnold
ART THEORY Cynthia Freeland
ARTIFICIAL INTELLIGENCE Margaret A. Boden
ASIAN AMERICAN HISTORY Madeline Y. Hsu
ASTROBIOLOGY David C. Catling
ASTROPHYSICS James Binney
ATHEISM Julian Baggini
THE ATMOSPHERE Paul I. Palmer
AUGUSTINE Henry Chadwick
JANE AUSTEN Tom Keymer
AUSTRALIA Kenneth Morgan
AUTISM Uta Frith
AUTOBIOGRAPHY Laura Marcus
THE AVANT GARDE David Cottington
THE AZTECS Davíd Carrasco
BABYLONIA Trevor Bryce
BACTERIA Sebastian G. B. Amyes
BANKING John Goddard and John O. S. Wilson
BARTHES Jonathan Culler
THE BEATS David Sterritt
BEAUTY Roger Scruton
LUDWIG VAN BEETHOVEN Mark Evan Bonds
BEHAVIOURAL ECONOMICS Michelle Baddeley
BESTSELLERS John Sutherland
THE BIBLE John Riches
BIBLICAL ARCHAEOLOGY Eric H. Cline
BIG DATA Dawn E. Holmes
BIOCHEMISTRY Mark Lorch
BIOGEOGRAPHY Mark V. Lomolino
BIOGRAPHY Hermione Lee
BIOMETRICS Michael Fairhurst
ELIZABETH BISHOP Jonathan F. S. Post
BLACK HOLES Katherine Blundell
BLASPHEMY Yvonne Sherwood
BLOOD Chris Cooper
THE BLUES Elijah Wald
THE BODY Chris Shilling
THE BOOK OF COMMON PRAYER Brian Cummings
THE BOOK OF MORMON Terryl Givens
BORDERS Alexander C. Diener and Joshua Hagen
THE BRAIN Michael O'Shea
BRANDING Robert Jones
THE BRICS Andrew F. Cooper
THE BRITISH CONSTITUTION Martin Loughlin
THE BRITISH EMPIRE Ashley Jackson
BRITISH POLITICS Tony Wright
BUDDHA Michael Carrithers
BUDDHISM Damien Keown
BUDDHIST ETHICS Damien Keown
BYZANTIUM Peter Sarris
CALVINISM Jon Balserak
ALBERT CAMUS Oliver Gloag
CANADA Donald Wright

CANCER Nicholas James
CAPITALISM James Fulcher
CATHOLICISM Gerald O'Collins
CAUSATION Stephen Mumford and Rani Lill Anjum
THE CELL Terence Allen and Graham Cowling
THE CELTS Barry Cunliffe
CHAOS Leonard Smith
GEOFFREY CHAUCER David Wallace
CHEMISTRY Peter Atkins
CHILD PSYCHOLOGY Usha Goswami
CHILDREN'S LITERATURE Kimberley Reynolds
CHINESE LITERATURE Sabina Knight
CHOICE THEORY Michael Allingham
CHRISTIAN ART Beth Williamson
CHRISTIAN ETHICS D. Stephen Long
CHRISTIANITY Linda Woodhead
CIRCADIAN RHYTHMS Russell Foster and Leon Kreitzman
CITIZENSHIP Richard Bellamy
CITY PLANNING Carl Abbott
CIVIL ENGINEERING David Muir Wood
CLASSICAL LITERATURE William Allan
CLASSICAL MYTHOLOGY Helen Morales
CLASSICS Mary Beard and John Henderson
CLAUSEWITZ Michael Howard
CLIMATE Mark Maslin
CLIMATE CHANGE Mark Maslin
CLINICAL PSYCHOLOGY Susan Llewelyn and Katie Aafjes-van Doorn
COGNITIVE BEHAVIOURAL THERAPY Freda McManus
COGNITIVE NEUROSCIENCE Richard Passingham
THE COLD WAR Robert J. McMahon
COLONIAL AMERICA Alan Taylor
COLONIAL LATIN AMERICAN LITERATURE Rolena Adorno
COMBINATORICS Robin Wilson
COMEDY Matthew Bevis
COMMUNISM Leslie Holmes
COMPARATIVE LITERATURE Ben Hutchinson
COMPETITION AND ANTITRUST LAW Ariel Ezrachi
COMPLEXITY John H. Holland
THE COMPUTER Darrel Ince
COMPUTER SCIENCE Subrata Dasgupta
CONCENTRATION CAMPS Dan Stone
CONFUCIANISM Daniel K. Gardner
THE CONQUISTADORS Matthew Restall and Felipe Fernández-Armesto
CONSCIENCE Paul Strohm
CONSCIOUSNESS Susan Blackmore
CONTEMPORARY ART Julian Stallabrass
CONTEMPORARY FICTION Robert Eaglestone
CONTINENTAL PHILOSOPHY Simon Critchley
COPERNICUS Owen Gingerich
CORAL REEFS Charles Sheppard
CORPORATE SOCIAL RESPONSIBILITY Jeremy Moon
CORRUPTION Leslie Holmes
COSMOLOGY Peter Coles
COUNTRY MUSIC Richard Carlin
CREATIVITY Vlad Glăveanu
CRIME FICTION Richard Bradford
CRIMINAL JUSTICE Julian V. Roberts
CRIMINOLOGY Tim Newburn
CRITICAL THEORY Stephen Eric Bronner
THE CRUSADES Christopher Tyerman
CRYPTOGRAPHY Fred Piper and Sean Murphy
CRYSTALLOGRAPHY A. M. Glazer
THE CULTURAL REVOLUTION Richard Curt Kraus
DADA AND SURREALISM David Hopkins
DANTE Peter Hainsworth and David Robey
DARWIN Jonathan Howard
THE DEAD SEA SCROLLS Timothy H. Lim
DECADENCE David Weir
DECOLONIZATION Dane Kennedy
DEMENTIA Kathleen Taylor
DEMOCRACY Bernard Crick

DEMOGRAPHY Sarah Harper
DEPRESSION Jan Scott and Mary Jane Tacchi
DERRIDA Simon Glendinning
DESCARTES Tom Sorell
DESERTS Nick Middleton
DESIGN John Heskett
DEVELOPMENT Ian Goldin
DEVELOPMENTAL BIOLOGY Lewis Wolpert
THE DEVIL Darren Oldridge
DIASPORA Kevin Kenny
CHARLES DICKENS Jenny Hartley
DICTIONARIES Lynda Mugglestone
DINOSAURS David Norman
DIPLOMATIC HISTORY Joseph M. Siracusa
DOCUMENTARY FILM Patricia Aufderheide
DREAMING J. Allan Hobson
DRUGS Les Iversen
DRUIDS Barry Cunliffe
DYNASTY Jeroen Duindam
DYSLEXIA Margaret J. Snowling
EARLY MUSIC Thomas Forrest Kelly
THE EARTH Martin Redfern
EARTH SYSTEM SCIENCE Tim Lenton
ECOLOGY Jaboury Ghazoul
ECONOMICS Partha Dasgupta
EDUCATION Gary Thomas
EGYPTIAN MYTH Geraldine Pinch
EIGHTEENTH-CENTURY BRITAIN Paul Langford
THE ELEMENTS Philip Ball
EMOTION Dylan Evans
EMPIRE Stephen Howe
EMPLOYMENT LAW David Cabrelli
ENERGY SYSTEMS Nick Jenkins
ENGELS Terrell Carver
ENGINEERING David Blockley
THE ENGLISH LANGUAGE Simon Horobin
ENGLISH LITERATURE Jonathan Bate
THE ENLIGHTENMENT John Robertson
ENTREPRENEURSHIP Paul Westhead and Mike Wright
ENVIRONMENTAL ECONOMICS Stephen Smith
ENVIRONMENTAL ETHICS Robin Attfield
ENVIRONMENTAL LAW Elizabeth Fisher
ENVIRONMENTAL POLITICS Andrew Dobson
ENZYMES Paul Engel
EPICUREANISM Catherine Wilson
EPIDEMIOLOGY Rodolfo Saracci
ETHICS Simon Blackburn
ETHNOMUSICOLOGY Timothy Rice
THE ETRUSCANS Christopher Smith
EUGENICS Philippa Levine
THE EUROPEAN UNION Simon Usherwood and John Pinder
EUROPEAN UNION LAW Anthony Arnull
EVANGELICALISM John Stackhouse
EVOLUTION Brian and Deborah Charlesworth
EXISTENTIALISM Thomas Flynn
EXPLORATION Stewart A. Weaver
EXTINCTION Paul B. Wignall
THE EYE Michael Land
FAIRY TALE Marina Warner
FAMILY LAW Jonathan Herring
MICHAEL FARADAY Frank A. J. L. James
FASCISM Kevin Passmore
FASHION Rebecca Arnold
FEDERALISM Mark J. Rozell and Clyde Wilcox
FEMINISM Margaret Walters
FILM Michael Wood
FILM MUSIC Kathryn Kalinak
FILM NOIR James Naremore
FIRE Andrew C. Scott
THE FIRST WORLD WAR Michael Howard
FLUID MECHANICS Eric Lauga
FOLK MUSIC Mark Slobin
FOOD John Krebs
FORENSIC PSYCHOLOGY David Canter
FORENSIC SCIENCE Jim Fraser
FORESTS Jaboury Ghazoul
FOSSILS Keith Thomson
FOUCAULT Gary Gutting
THE FOUNDING FATHERS R. B. Bernstein

FRACTALS Kenneth Falconer
FREE SPEECH Nigel Warburton
FREE WILL Thomas Pink
FREEMASONRY Andreas Önnerfors
FRENCH LITERATURE John D. Lyons
FRENCH PHILOSOPHY Stephen Gaukroger and Knox Peden
THE FRENCH REVOLUTION William Doyle
FREUD Anthony Storr
FUNDAMENTALISM Malise Ruthven
FUNGI Nicholas P. Money
THE FUTURE Jennifer M. Gidley
GALAXIES John Gribbin
GALILEO Stillman Drake
GAME THEORY Ken Binmore
GANDHI Bhikhu Parekh
GARDEN HISTORY Gordon Campbell
GENES Jonathan Slack
GENIUS Andrew Robinson
GENOMICS John Archibald
GEOGRAPHY John Matthews and David Herbert
GEOLOGY Jan Zalasiewicz
GEOMETRY Maciej Dunajski
GEOPHYSICS William Lowrie
GEOPOLITICS Klaus Dodds
GERMAN LITERATURE Nicholas Boyle
GERMAN PHILOSOPHY Andrew Bowie
THE GHETTO Bryan Cheyette
GLACIATION David J. A. Evans
GLOBAL CATASTROPHES Bill McGuire
GLOBAL ECONOMIC HISTORY Robert C. Allen
GLOBAL ISLAM Nile Green
GLOBALIZATION Manfred B. Steger
GOD John Bowker
GOETHE Ritchie Robertson
THE GOTHIC Nick Groom
GOVERNANCE Mark Bevir
GRAVITY Timothy Clifton
THE GREAT DEPRESSION AND THE NEW DEAL Eric Rauchway
HABEAS CORPUS Amanda Tyler
HABERMAS James Gordon Finlayson
THE HABSBURG EMPIRE Martyn Rady
HAPPINESS Daniel M. Haybron
THE HARLEM RENAISSANCE Cheryl A. Wall
THE HEBREW BIBLE AS LITERATURE Tod Linafelt
HEGEL Peter Singer
HEIDEGGER Michael Inwood
THE HELLENISTIC AGE Peter Thonemann
HEREDITY John Waller
HERMENEUTICS Jens Zimmermann
HERODOTUS Jennifer T. Roberts
HIEROGLYPHS Penelope Wilson
HINDUISM Kim Knott
HISTORY John H. Arnold
THE HISTORY OF ASTRONOMY Michael Hoskin
THE HISTORY OF CHEMISTRY William H. Brock
THE HISTORY OF CHILDHOOD James Marten
THE HISTORY OF CINEMA Geoffrey Nowell-Smith
THE HISTORY OF LIFE Michael Benton
THE HISTORY OF MATHEMATICS Jacqueline Stedall
THE HISTORY OF MEDICINE William Bynum
THE HISTORY OF PHYSICS J. L. Heilbron
THE HISTORY OF POLITICAL THOUGHT Richard Whatmore
THE HISTORY OF TIME Leofranc Holford Strevens
HIV AND AIDS Alan Whiteside
HOBBES Richard Tuck
HOLLYWOOD Peter Decherney
THE HOLY ROMAN EMPIRE Joachim Whaley
HOME Michael Allen Fox
HOMER Barbara Graziosi
HORMONES Martin Luck
HORROR Darryl Jones
HUMAN ANATOMY Leslie Klenerman
HUMAN EVOLUTION Bernard Wood
HUMAN PHYSIOLOGY Jamie A. Davies

HUMAN RESOURCE MANAGEMENT Adrian Wilkinson
HUMAN RIGHTS Andrew Clapham
HUMANISM Stephen Law
HUME James A. Harris
HUMOUR NOËL Carroll
THE ICE AGE Jamie Woodward
IDENTITY Florian Coulmas
IDEOLOGY Michael Freeden
THE IMMUNE SYSTEM Paul Klenerman
INDIAN CINEMA Ashish Rajadhyaksha
INDIAN PHILOSOPHY Sue Hamilton
THE INDUSTRIAL REVOLUTION Robert C. Allen
INFECTIOUS DISEASE Marta L. Wayne and Benjamin M. Bolker
INFINITY Ian Stewart
INFORMATION Luciano Floridi
INNOVATION Mark Dodgson and David Gann
INTELLECTUAL PROPERTY Siva Vaidhyanathan
INTELLIGENCE Ian J. Deary
INTERNATIONAL LAW Vaughan Lowe
INTERNATIONAL MIGRATION Khalid Koser
INTERNATIONAL RELATIONS Christian Reus-Smit
INTERNATIONAL SECURITY Christopher S. Browning
IRAN Ali M. Ansari
ISLAM Malise Ruthven
ISLAMIC HISTORY Adam Silverstein
ISLAMIC LAW Mashood A. Baderin
ISOTOPES Rob Ellam
ITALIAN LITERATURE Peter Hainsworth and David Robey
HENRY JAMES Susan L. Mizruchi
JESUS Richard Bauckham
JEWISH HISTORY David N. Myers
JEWISH LITERATURE Ilan Stavans
JOURNALISM Ian Hargreaves
JAMES JOYCE Colin MacCabe
JUDAISM Norman Solomon
JUNG Anthony Stevens
KABBALAH Joseph Dan
KAFKA Ritchie Robertson
KANT Roger Scruton
KEYNES Robert Skidelsky
KIERKEGAARD Patrick Gardiner
KNOWLEDGE Jennifer Nagel
THE KORAN Michael Cook
KOREA Michael J. Seth
LAKES Warwick F. Vincent
LANDSCAPE ARCHITECTURE Ian H. Thompson
LANDSCAPES AND GEOMORPHOLOGY Andrew Goudie and Heather Viles
LANGUAGES Stephen R. Anderson
LATE ANTIQUITY Gillian Clark
LAW Raymond Wacks
THE LAWS OF THERMODYNAMICS Peter Atkins
LEADERSHIP Keith Grint
LEARNING Mark Haselgrove
LEIBNIZ Maria Rosa Antognazza
C. S. LEWIS James Como
LIBERALISM Michael Freeden
LIGHT Ian Walmsley
LINCOLN Allen C. Guelzo
LINGUISTICS Peter Matthews
LITERARY THEORY Jonathan Culler
LOCKE John Dunn
LOGIC Graham Priest
LOVE Ronald de Sousa
MARTIN LUTHER Scott H. Hendrix
MACHIAVELLI Quentin Skinner
MADNESS Andrew Scull
MAGIC Owen Davies
MAGNA CARTA Nicholas Vincent
MAGNETISM Stephen Blundell
MALTHUS Donald Winch
MAMMALS T. S. Kemp
MANAGEMENT John Hendry
NELSON MANDELA Elleke Boehmer
MAO Delia Davin
MARINE BIOLOGY Philip V. Mladenov
MARKETING Kenneth Le Meunier-FitzHugh
THE MARQUIS DE SADE John Phillips
MARTYRDOM Jolyon Mitchell
MARX Peter Singer
MATERIALS Christopher Hall

MATHEMATICAL FINANCE Mark H. A. Davis
MATHEMATICS Timothy Gowers
MATTER Geoff Cottrell
THE MAYA Matthew Restall and Amara Solari
THE MEANING OF LIFE Terry Eagleton
MEASUREMENT David Hand
MEDICAL ETHICS Michael Dunn and Tony Hope
MEDICAL LAW Charles Foster
MEDIEVAL BRITAIN John Gillingham and Ralph A. Griffiths
MEDIEVAL LITERATURE Elaine Treharne
MEDIEVAL PHILOSOPHY John Marenbon
MEMORY Jonathan K. Foster
METAPHYSICS Stephen Mumford
METHODISM William J. Abraham
THE MEXICAN REVOLUTION Alan Knight
MICROBIOLOGY Nicholas P. Money
MICROECONOMICS Avinash Dixit
MICROSCOPY Terence Allen
THE MIDDLE AGES Miri Rubin
MILITARY JUSTICE Eugene R. Fidell
MILITARY STRATEGY Antulio J. Echevarria II
JOHN STUART MILL Gregory Claeys
MINERALS David Vaughan
MIRACLES Yujin Nagasawa
MODERN ARCHITECTURE Adam Sharr
MODERN ART David Cottington
MODERN BRAZIL Anthony W. Pereira
MODERN CHINA Rana Mitter
MODERN DRAMA Kirsten E. Shepherd-Barr
MODERN FRANCE Vanessa R. Schwartz
MODERN INDIA Craig Jeffrey
MODERN IRELAND Senia Pašeta
MODERN ITALY Anna Cento Bull
MODERN JAPAN Christopher Goto-Jones
MODERN LATIN AMERICAN LITERATURE Roberto González Echevarría
MODERN WAR Richard English
MODERNISM Christopher Butler
MOLECULAR BIOLOGY Aysha Divan and Janice A. Royds
MOLECULES Philip Ball
MONASTICISM Stephen J. Davis
THE MONGOLS Morris Rossabi
MONTAIGNE William M. Hamlin
MOONS David A. Rothery
MORMONISM Richard Lyman Bushman
MOUNTAINS Martin F. Price
MUHAMMAD Jonathan A. C. Brown
MULTICULTURALISM Ali Rattansi
MULTILINGUALISM John C. Maher
MUSIC Nicholas Cook
MYTH Robert A. Segal
NAPOLEON David Bell
THE NAPOLEONIC WARS Mike Rapport
NATIONALISM Steven Grosby
NATIVE AMERICAN LITERATURE Sean Teuton
NAVIGATION Jim Bennett
NAZI GERMANY Jane Caplan
NEOLIBERALISM Manfred B. Steger and Ravi K. Roy
NETWORKS Guido Caldarelli and Michele Catanzaro
THE NEW TESTAMENT Luke Timothy Johnson
THE NEW TESTAMENT AS LITERATURE Kyle Keefer
NEWTON Robert Iliffe
NIELS BOHR J. L. Heilbron
NIETZSCHE Michael Tanner
NINETEENTH-CENTURY BRITAIN Christopher Harvie and H. C. G. Matthew
THE NORMAN CONQUEST George Garnett
NORTH AMERICAN INDIANS Theda Perdue and Michael D. Green
NORTHERN IRELAND Marc Mulholland
NOTHING Frank Close
NUCLEAR PHYSICS Frank Close
NUCLEAR POWER Maxwell Irvine
NUCLEAR WEAPONS Joseph M. Siracusa

NUMBER THEORY Robin Wilson
NUMBERS Peter M. Higgins
NUTRITION David A. Bender
OBJECTIVITY Stephen Gaukroger
OCEANS Dorrik Stow
THE OLD TESTAMENT Michael D. Coogan
THE ORCHESTRA D. Kern Holoman
ORGANIC CHEMISTRY Graham Patrick
ORGANIZATIONS Mary Jo Hatch
ORGANIZED CRIME Georgios A. Antonopoulos and Georgios Papanicolaou
ORTHODOX CHRISTIANITY A. Edward Siecienski
OVID Llewelyn Morgan
PAGANISM Owen Davies
PAKISTAN Pippa Virdee
THE PALESTINIAN-ISRAELI CONFLICT Martin Bunton
PANDEMICS Christian W. McMillen
PARTICLE PHYSICS Frank Close
PAUL E. P. Sanders
PEACE OLIVER P. Richmond
PENTECOSTALISM William K. Kay
PERCEPTION Brian Rogers
THE PERIODIC TABLE Eric R. Scerri
PHILOSOPHICAL METHOD Timothy Williamson
PHILOSOPHY Edward Craig
PHILOSOPHY IN THE ISLAMIC WORLD Peter Adamson
PHILOSOPHY OF BIOLOGY Samir Okasha
PHILOSOPHY OF LAW Raymond Wacks
PHILOSOPHY OF MIND Barbara Gail Montero
PHILOSOPHY OF PHYSICS David Wallace
PHILOSOPHY OF SCIENCE Samir Okasha
PHILOSOPHY OF RELIGION Tim Bayne
PHOTOGRAPHY Steve Edwards
PHYSICAL CHEMISTRY Peter Atkins
PHYSICS Sidney Perkowitz
PILGRIMAGE Ian Reader
PLAGUE Paul Slack
PLANETARY SYSTEMS Raymond T. Pierrehumbert
PLANETS David A. Rothery
PLANTS Timothy Walker
PLATE TECTONICS Peter Molnar
PLATO Julia Annas
POETRY Bernard O'Donoghue
POLITICAL PHILOSOPHY David Miller
POLITICS Kenneth Minogue
POLYGAMY Sarah M. S. Pearsall
POPULISM Cas Mudde and Cristóbal Rovira Kaltwasser
POSTCOLONIALISM Robert Young
POSTMODERNISM Christopher Butler
POSTSTRUCTURALISM Catherine Belsey
POVERTY Philip N. Jefferson
PREHISTORY Chris Gosden
PRESOCRATIC PHILOSOPHY Catherine Osborne
PRIVACY Raymond Wacks
PROBABILITY John Haigh
PROGRESSIVISM Walter Nugent
PROHIBITION W. J. Rorabaugh
PROJECTS Andrew Davies
PROTESTANTISM Mark A. Noll
PSYCHIATRY Tom Burns
PSYCHOANALYSIS Daniel Pick
PSYCHOLOGY Gillian Butler and Freda McManus
PSYCHOLOGY OF MUSIC Elizabeth Hellmuth Margulis
PSYCHOPATHY Essi Viding
PSYCHOTHERAPY Tom Burns and Eva Burns-Lundgren
PUBLIC ADMINISTRATION Stella Z. Theodoulou and Ravi K. Roy
PUBLIC HEALTH Virginia Berridge
PURITANISM Francis J. Bremer
THE QUAKERS Pink Dandelion
QUANTUM THEORY John Polkinghorne
RACISM Ali Rattansi
RADIOACTIVITY Claudio Tuniz
RASTAFARI Ennis B. Edmonds
READING Belinda Jack

THE REAGAN REVOLUTION Gil Troy
REALITY Jan Westerhoff
RECONSTRUCTION Allen C. Guelzo
THE REFORMATION Peter Marshall
REFUGEES Gil Loescher
RELATIVITY Russell Stannard
RELIGION Thomas A. Tweed
RELIGION IN AMERICA Timothy Beal
THE RENAISSANCE Jerry Brotton
RENAISSANCE ART Geraldine A. Johnson
RENEWABLE ENERGY Nick Jelley
REPTILES T. S. Kemp
REVOLUTIONS Jack A. Goldstone
RHETORIC Richard Toye
RISK Baruch Fischhoff and John Kadvany
RITUAL Barry Stephenson
RIVERS Nick Middleton
ROBOTICS Alan Winfield
ROCKS Jan Zalasiewicz
ROMAN BRITAIN Peter Salway
THE ROMAN EMPIRE Christopher Kelly
THE ROMAN REPUBLIC David M. Gwynn
ROMANTICISM Michael Ferber
ROUSSEAU Robert Wokler
RUSSELL A. C. Grayling
THE RUSSIAN ECONOMY Richard Connolly
RUSSIAN HISTORY Geoffrey Hosking
RUSSIAN LITERATURE Catriona Kelly
THE RUSSIAN REVOLUTION S. A. Smith
SAINTS Simon Yarrow
SAMURAI Michael Wert
SAVANNAS Peter A. Furley
SCEPTICISM Duncan Pritchard
SCHIZOPHRENIA Chris Frith and Eve Johnstone
SCHOPENHAUER Christopher Janaway
SCIENCE AND RELIGION Thomas Dixon and Adam R. Shapiro
SCIENCE FICTION David Seed
THE SCIENTIFIC REVOLUTION Lawrence M. Principe
SCOTLAND Rab Houston
SECULARISM Andrew Copson
SEXUAL SELECTION Marlene Zuk and Leigh W. Simmons
SEXUALITY Véronique Mottier
WILLIAM SHAKESPEARE Stanley Wells
SHAKESPEARE'S COMEDIES Bart van Es
SHAKESPEARE'S SONNETS AND POEMS Jonathan F. S. Post
SHAKESPEARE'S TRAGEDIES Stanley Wells
GEORGE BERNARD SHAW Christopher Wixson
MARY SHELLEY Charlotte Gordon
THE SHORT STORY Andrew Kahn
SIKHISM Eleanor Nesbitt
SILENT FILM Donna Kornhaber
THE SILK ROAD James A. Millward
SLANG Jonathon Green
SLEEP Steven W. Lockley and Russell G. Foster
SMELL Matthew Cobb
ADAM SMITH Christopher J. Berry
SOCIAL AND CULTURAL ANTHROPOLOGY John Monaghan and Peter Just
SOCIAL PSYCHOLOGY Richard J. Crisp
SOCIAL WORK Sally Holland and Jonathan Scourfield
SOCIALISM Michael Newman
SOCIOLINGUISTICS John Edwards
SOCIOLOGY Steve Bruce
SOCRATES C. C. W. Taylor
SOFT MATTER Tom McLeish
SOUND Mike Goldsmith
SOUTHEAST ASIA James R. Rush
THE SOVIET UNION Stephen Lovell
THE SPANISH CIVIL WAR Helen Graham
SPANISH LITERATURE Jo Labanyi
THE SPARTANS Andrew J. Bayliss
SPINOZA Roger Scruton

SPIRITUALITY Philip Sheldrake
SPORT Mike Cronin
STARS Andrew King
STATISTICS David J. Hand
STEM CELLS Jonathan Slack
STOICISM Brad Inwood
STRUCTURAL ENGINEERING David Blockley
STUART BRITAIN John Morrill
THE SUN Philip Judge
SUPERCONDUCTIVITY Stephen Blundell
SUPERSTITION Stuart Vyse
SYMMETRY Ian Stewart
SYNAESTHESIA Julia Simner
SYNTHETIC BIOLOGY Jamie A. Davies
SYSTEMS BIOLOGY Eberhard O. Voit
TAXATION Stephen Smith
TEETH Peter S. Ungar
TELESCOPES Geoff Cottrell
TERRORISM Charles Townshend
THEATRE Marvin Carlson
THEOLOGY David F. Ford
THINKING AND REASONING Jonathan St B. T. Evans
THOUGHT Tim Bayne
TIBETAN BUDDHISM Matthew T. Kapstein
TIDES David George Bowers and Emyr Martyn Roberts
TIME Jenann Ismael
TOCQUEVILLE Harvey C. Mansfield
LEO TOLSTOY Liza Knapp
TOPOLOGY Richard Earl
TRAGEDY Adrian Poole
TRANSLATION Matthew Reynolds
THE TREATY OF VERSAILLES Michael S. Neiberg
TRIGONOMETRY Glen Van Brummelen
THE TROJAN WAR Eric H. Cline
TRUST Katherine Hawley
THE TUDORS John Guy
TWENTIETH-CENTURY BRITAIN Kenneth O. Morgan
TYPOGRAPHY Paul Luna
THE UNITED NATIONS Jussi M. Hanhimäki
UNIVERSITIES AND COLLEGES David Palfreyman and Paul Temple
THE U.S. CIVIL WAR Louis P. Masur
THE U.S. CONGRESS Donald A. Ritchie
THE U.S. CONSTITUTION David J. Bodenhamer
THE U.S. SUPREME COURT Linda Greenhouse
UTILITARIANISM Katarzyna de Lazari-Radek and Peter Singer
UTOPIANISM Lyman Tower Sargent
VETERINARY SCIENCE James Yeates
THE VIKINGS Julian D. Richards
VIOLENCE Philip Dwyer
THE VIRGIN MARY Mary Joan Winn Leith
THE VIRTUES Craig A. Boyd and Kevin Timpe
VIRUSES Dorothy H. Crawford
VOLCANOES Michael J. Branney and Jan Zalasiewicz
VOLTAIRE Nicholas Cronk
WAR AND RELIGION Jolyon Mitchell and Joshua Rey
WAR AND TECHNOLOGY Alex Roland
WATER John Finney
WAVES Mike Goldsmith
WEATHER Storm Dunlop
THE WELFARE STATE David Garland
WITCHCRAFT Malcolm Gaskill
WITTGENSTEIN A. C. Grayling
WORK Stephen Fineman
WORLD MUSIC Philip Bohlman
THE WORLD TRADE ORGANIZATION Amrita Narlikar
WORLD WAR II Gerhard L. Weinberg
WRITING AND SCRIPT Andrew Robinson
ZIONISM Michael Stanislawski
ÉMILE ZOLA Brian Nelson

Available soon:

HANNAH ARENDT Dana Villa
INSECTS Simon R. Leather
BUSINESS ETHICS Nils Ole Oermann
BRITISH CINEMA Charles Barr
THE NOVEL Robert Eaglestone

For more information visit our website
www.oup.com/vsi/

Eric Lauga

FLUID MECHANICS

A Very Short Introduction

OXFORD
UNIVERSITY PRESS

Great Clarendon Street, Oxford, OX2 6DP,
United Kingdom

Oxford University Press is a department of the University of Oxford.
It furthers the University's objective of excellence in research, scholarship,
and education by publishing worldwide. Oxford is a registered trade mark of
Oxford University Press in the UK and in certain other countries

First edition published in 2022

Published in the United States of America by Oxford University Press
198 Madison Avenue, New York, NY 10016, United States of America

British Library Cataloguing in Publication Data
Data available

Library of Congress Control Number: 2021953271

ISBN 978–0–19–883100–6

Printed and bound by CPI Group (UK) Ltd, Croydon, CR0 4YY

Contents

Preface

Fluid mechanics is many things to many people.

On one hand, it is the study of fluid motion, i.e. how liquids and gases flow when we push them and how they push us back. But since flows affect many areas of science and engineering, fluid mechanics is also the use of this knowledge to help answer scientific questions that impact our daily lives. As a result, fluid mechanics is often combined with other scientific branches into one that is both all-encompassing and hard to define.

This ever-changing identity of fluid mechanics is reflected in how it is taught at university. Unlike linear algebra or quantum mechanics, no consensus exists on who exactly should teach it. In some institutions, fluid mechanics is part of mathematics, while in others it is engineering or physics. Like a nomadic science, fluid mechanics belongs nowhere but can find a home anywhere.

Given its interdisciplinary nature, it is inevitable that students from many different backgrounds should be exposed to it. Learning fluid mechanics can thus take different approaches, from one with the full mathematical complexity, to an applied point of view emphasizing its applications, or a more physical one emphasizing fundamental mechanisms. All of these approaches are of course valid and depend on the perspective. Taking a bath and wondering about buoyancy does not require the same level of knowledge as building aeroplanes or designing oil pipes.

Fluid mechanics is often perceived as being mathematically daunting. In this book, I present a short introduction to the field by focusing on the fundamental physical ideas. My main goal is to communicate the intuition behind a wide range of flow behaviours in a manner requiring minimal mathematical knowledge. By the end of the book, I hope that readers will have developed new understanding along four strands: an appreciation for the intrinsic beauty of the field; a new (or renewed) physical intuition for fluid motion; an appreciation for the impact of flows on other scientific fields; and an awareness of the role of fluid motion in both the natural and industrial world.

Many thanks to Alain Goriely for early encouragements, to Weida Liao for her kind feedback on every chapter, and to my wife Dominique for a detailed read of the manuscript and overall moral support. Biographical sketches are adapted in part from online resources.

Cambridge, UK
October 2021

Eric Lauga

List of illustrations

Chapter 1
Fluids

The scientific study of fluid motion—fluid mechanics—has long been motivated by two purposes. We all want to understand the world around us, and since our world is dominated by the presence of fluids, it is full of fundamental questions involving fluid motion. How can aeroplanes fly? Why do waves break? Why doesn't water wet lotus leaves?

Fluid flows also affect many aspects of the industrial world. There is therefore an ongoing commercial need to quantify fluid behaviour, predict it, and manage the interactions between fluids and their environments. Practical questions abound for which fluid mechanics is the key. Can we reduce drag on aeroplanes? How should surfboards be designed? Can we use lotus-leaf technology to create waterproof clothes?

In this first chapter, we start by considering the structure of fluids and discuss some of the remarkable properties they display even in the absence of flow.

Fluids and molecules

The first logical question to ask is: what exactly is a fluid? One natural way to answer this question is to look in detail at the molecular structure of the three familiar states of matter. Both liquids and gases are fluids. In the solid state, individual particles (be they atoms, molecules, or even charged particles called ions) are held together by strong forces and arranged in a very regular symmetrical pattern called a crystal, which is periodically repeated over the entire length of the solid. The individual particles in a crystal cannot move, and this explains why solids have a rigid structure.

In contrast, particles in liquids and gases can move freely. In the case of liquids, particles are very close to each other, but they are constantly jiggling around, and may rearrange in response to external forces. The molecular structure of liquids therefore allows them to flow. The most fundamental of all liquids is, of course, water, while blood is an example of a complex liquid vital to animal life (Figure 1).

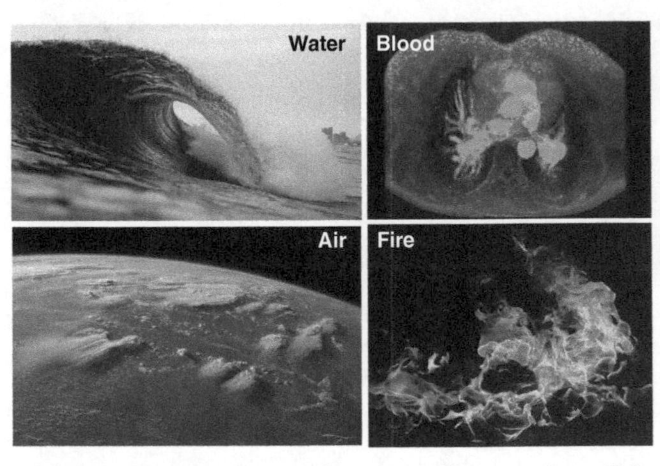

1. Four examples of fluid motion illustrating their impact in our daily life. An ocean water wave in the process of breaking is an example of a simple liquid. Blood flow in an aorta is an example of a complex liquid (levels of grey measure the average flow speed). Cumulonimbus clouds in the stratosphere near Borneo, Indonesia, are examples of a flowing gas. A fire is an example of complex flow of gas undergoing chemical reactions.

The molecular structure of gases is very different from that of liquids, and yet gases are also fluids. In a gas, particles are widely separated and almost don't interact with one another except for occasional collisions. But particles are able to move in response to forces and therefore gases do flow. The air in our surrounding atmosphere hides in plain sight but clouds can be used to visualize its motion, while fire is an example of a gas flowing while undergoing chemical reactions (Figure 1).

The stark difference between the molecular structure of liquids and gases has some important consequences. For example, it is much easier to compress a gas than it is to compress a liquid, since gases have a lot of free space between particles whereas liquids have almost none. Solids and liquids also tend to be very similar in weight, whereas gases are of course much lighter. So one could argue that, from a molecular standpoint, liquids very much resemble solids. But in terms of their ability to deform in response to external forces and to flow, liquids and gases are very similar, and both are fluids.

Macroscopic fluids

The problem with a molecular answer to the definition of a fluid is that it is not very satisfying. In our everyday experience, most of us do not apprehend reality with molecular glasses, so there must be a way to define a fluid without having to resort to molecular concepts. Counter-intuitively, it is easier to define what is *not* a fluid. The opposite of a fluid is called a solid, a material that has a fixed shape and is able to keep it permanently. A solid can often deform if an external force is applied to it, but when the force is removed the solid returns to its original shape with no permanent change to its state. Picture a trampoline or a sponge: you can try as hard as you can to deform them, they will always bounce back.

A fluid is what is not a solid, and therefore it has the opposite characteristics. A fluid does not have a fixed shape but constantly changes its shape in order to adapt to the form of its surroundings. A fluid flows continuously under the action of an imposed force, and this deformation is permanent with no memory of where it started. Both of these statements apply equally to liquids (for example, water in the ocean or in a cup of tea) and gases (for example, the air surrounding us). Thus both liquids and gases are fluids.

Alternatively, scientists like to say that fluids are defined as materials that cannot sustain shear forces. While we will focus on shear in Chapter 2, the intuitive way to think about it is to imagine depositing a drop of saliva between your thumb and another finger. If you slide one finger relative to the other, the saliva flows continuously and, in fact, if you had infinite fingers and infinite saliva, you could do this forever. In contrast, if you deposit a small piece of wood or a mobile phone between your fingers and try to slide without slipping, the material resists the deformation. Solids cannot be sheared but fluids can.

As with all scientific classification, exceptions exist. The distinction outlined above separates simple solids from simple fluids. In contrast, complex fluids are materials that share some characteristics with both fluids and solids. For example, shaving foams or gels can flow, but they retain some memory of their initial state. Similarly, toothpaste has a fixed shape when deposited on a toothbrush and yet it offers very little resistance to being rubbed against our teeth.

The continuum hypothesis

When we look at a table, we don't see the molecules that constitute it and their continuous vibrations. Instead, we see a piece of material that is uniform and at rest. The same is true for water in a glass, which fills all of the space available to it in a continuous fashion. Yet, we know that vibrating particles are the fundamental constituents of matter. With a sufficiently powerful zoom, every material would appear to us as made up of discrete elements constantly moving around and interacting with one another. How is it possible that a fluid can then both be made of many distinct elements and yet appear to us as continuous and smooth?

The response to this lies in the concept of averaging. To gain intuition, let us first consider the case of sand. The prototypical example of a granular material, sand is made up of individual sand grains. Each grain has a slightly different shape, a different colour, and originates from a slightly different rock and mineral. The sand is also being moved around by wind or by children running on it. Yet when we drive up to the beach and see the sand dunes in the horizon, the same sand appears to us to be completely uniform and smooth, and we don't see its motion at all. The difference between holding sand in your hand and seeing sand from your car is therefore one of length scale. In the first case you are looking at the sand a few centimetres away from it and can resolve the motion of its grains, while in the other case you are kilometres away from it and every little sand movement is averaging out to zero. So depending on the length scale at which we appreciate the material, it can appear to be spatially discrete and moving around, or spatially continuous and at rest.

What is true for sand is similarly true for a fluid. Molecules in the glass of water constantly move around and collide with one another. But if we zoom out, we average their motion on length scales much larger than the distance between the molecules, and obtain zero net motion and a continuous material. Scientists call these types of materials a continuum, and the branch of physics devoted to quantifying how they deform in response to forces is known as continuum mechanics.

A continuum fluid, the one that is intuitively known to us, is thus one obtained by ignoring its discrete nature and by averaging the motion of its molecules with that of its neighbours. How exactly should this averaging be done? If we average over too few neighbours, the fluid motion will still include some trace of the molecular randomness. At the other end, if we average over too many molecules, we will not be able to resolve the well-defined length scales that characterize fluid motion. For example, to describe water waves crashing on a beach, we need to capture the typical length scales between each wave, their typical height but also the fine features of the water droplets that are produced when waves break.

Scientists have been able to probe this question using controlled experiments on minute amounts of fluids. Simple liquids such as water behave as a smooth continuum as soon as the length scales in the fluid exceed a few nanometres (it takes 10^9 nanometres to make a metre). Since the typical molecular size is a tenth of a nanometre, it means that averaging motion over a few tens of individual particles gives rise to a continuum liquid. The situation in gases is similar but, since the typical distance between gas molecules is much larger than in liquids, the averaging length scales are larger. For example, gas molecules in air travel on average 50 nanometres between successive collisions. The continuum approximation tends therefore to be valid for gases on length scales of micrometres or more (it takes 10^6 micrometres to make a metre and thus 10^3 nanometres to make a micrometre).

Notwithstanding exceptions, such as the upper parts of the atmosphere where air is rarefied and gas molecules have to travel much more between collisions, all fluids on human length scales (say, millimetres or more) are safely in the continuum limit. This is the limit we will focus on in this book. In particular, it means that we will be able to refer to the properties of the fluid at a particular point of space and time, for example its velocity, with the implicit assumption that it means the local value averaged over a few tens of fluid particles.

Pressure

Fluid mechanics is mostly focused on the study of fluids that are deformed and flow, and as a result the field is often called fluid dynamics. But even in the absence of motion, fluids have remarkable characteristics. When they do not move, fluids are said to be at equilibrium and their study is that of fluid statics. One of the most important properties of fluids at equilibrium is their ability to sustain, and to exert, pressure.

Imagine putting a volume of a liquid in a large container and covering the fluid with a tight-fitting lid (i.e. one that does not allow any fluid to escape on the sides). If it wasn't for the fluid, the lid would be free to slide up and down the sides of the container like a piston. If you then place a heavy weight on the lid, this weight exerts a force on the fluid, represented by the black arrow pointing downward in Figure 2(a). If the fluid is a liquid, the presence of the weight leads to no visible change because liquids are essentially incompressible. Since the lid does not move at all, it means that the sum of the forces acting on it must be zero. Consequently, a force equal and opposite to the external force acting on the lid must be applied by the liquid, as represented by the grey upward arrow in Figure 2(a). We have thus uncovered a remarkable property of a liquid: if you push on it, it is able to push back by exactly the same amount. Push harder, and the fluid will also push harder.

The ability of a liquid at rest to generate a force in response to an external push is referred to as pressure. The value of the external force acting in Figure 2(a) divided by the surface area over which the force is applied gives a number called the external pressure. In response to this external pressure, liquids have the ability to generate an identical internal pressure, whatever its value. In fact, this is also true if the external pressure is negative, i.e. if the direction of the external force in Figure 2(a) is reversed. If you pull on a liquid, the liquid will pull back by the same amount, and it is then said to be under tension.

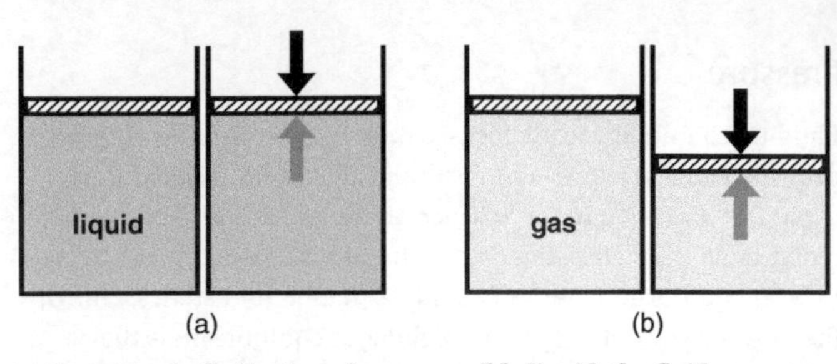

2. Pressures in fluids. In an incompressible liquid, the fluid pressure adapts to the value of the pressure generated by an external force applied to it (a). In contrast, a compressible gas will decrease its volume in response to an external force until the pressure in the gas balances the external pressure (b).

If instead of a liquid, the fluid in the container is a gas, the behaviour is slightly different. Consider first an open container. The air in the container is composed of moving particles and they collide frequently with the walls of the container, exerting a force on them. The value of this force, divided by the surface area on which it acts, is the pressure in the gas. Incidentally, this is the same value you would measure in a liquid placed in the container but with no lid on, and is due to the external pressure generated by the collisions between the surrounding air and the surface of the liquid.

Now, imagine placing the tight-fitting lid on the air-filled container as we did in the case of a liquid, and applying an external downward force to it, for example by using a weight (Figure 2(b)). In contrast with liquids, gases are compressible and therefore the lid moves down a small amount. As the lid slides down, the volume available to the air trapped in the container decreases so the particles in the closed container have a bit less space and thus collide more often with the sides of the container. This in turn means that the pressure in the container increases.

As the lid continues to move down, at some point the increase in gas pressure will be sufficiently large that the force applied by the gas in the container on the lid will be equal to the external force. At that point, the forces on the lid balance and it stops moving. The end result is similar to the case of a liquid: if you push on a gas, it is eventually able to push back by the same amount, with the difference that the gas gets compressed in the process so as to balance the external forces applied to it.

In addition to being able to adapt in response to external conditions, the pressure in a fluid has another important property. No matter which direction you look, the pressure has the same value. Isotropy is the term used to describe this property, and the pressure in the fluid is said to be isotropic. We can describe this feature using another thought experiment. Imagine quantifying the value of the pressure in the compressed gas of Figure 2(b) by placing a small flat probe inside the container and measuring the force exerted by gas molecules colliding with it. The measured pressure is equal to the value of this force divided by the area of the probe. If you place the probe at one point, measure the pressure, and then rotate the probe in place and re-do the measurement, you will obtain the same pressure value. Repeat with any rotation in three-dimensional space and the measurement for the pressure is always the same. Pressure acts, therefore, in all directions with the same magnitude.

This property of pressure isotropy is called the law of pressure and is originally due to the work of the 17th-century French mathematician Blaise Pascal. Today, Pascal is mostly known to the general public for his work on philosophy and in particular his scientific argument for believing in God. Pascal's last name is now used as the unit of pressure, with 1 Pascal of pressure (1 Pa) equal to a weight of 1 kilogram acting over 1 square metre. Pressure in the air surrounding us under normal room-temperature conditions and at sea level is approximately 100,000 Pa.

Hydrostatics

We discussed how the pressure in a fluid can result from an externally applied force (Figure 2). A lid with a weight on it pushes on the fluid, which pushes back. But the force generating the fluid pressure does not have to be externally imposed, and in fact it can arise from the fluid itself. Indeed fluids, just like everything around us, are subject to the unavoidable force of gravity.

Gravity is one of the fundamental forces of nature. As originally derived by the 17th-century English mathematician Isaac Newton, one of the most influential scientists of all time, two objects with a mass always attract each other with a gravitational force proportional to the product of their masses. On Earth, a body of mass m (small compared to that of the Earth) experiences a downward force due its gravitational attraction by Earth called the weight W, of magnitude $W = mg$, where g is the acceleration due to gravity (approximately a constant value on Earth).

Applied to a fluid, the presence of this gravitational force means that each small parcel of fluid experiences its own downward force, with a rather dramatic consequence for the state of pressure in the fluid. Consider, for example, a liquid in a container and imagine drawing an invisible horizontal surface somewhere in it (Figure 3(a)). The fluid located between the top surface and the invisible horizontal surface is at equilibrium so the sum of all the vertical forces acting on it must be zero. That fluid is subject to two downward forces: the total weight of the liquid plus the force from the external air pressure. Balancing these two downward forces is the upward force from the fluid pressure immediately underneath the invisible horizontal surface.

The pressure in the liquid therefore needs to balance the weight. However, the value of the weight depends on where we draw our invisible horizontal surface. If the surface is drawn near the top,

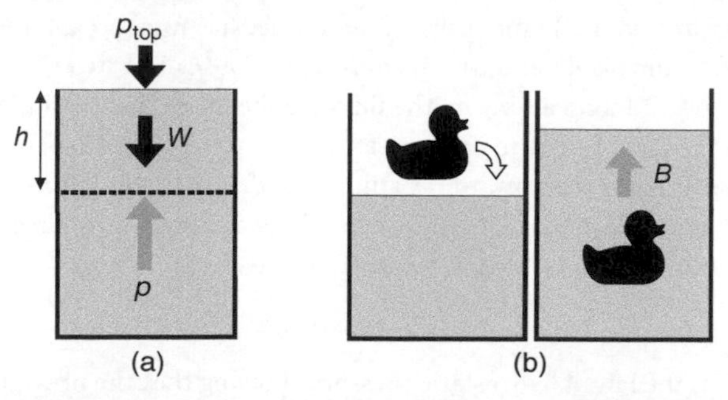

3. Hydrostatic pressure and buoyancy in a liquid. A portion of fluid of depth *h* is subject to two downward forces (a), its weight *W* and the external pressure p_{top} above its top surface, and to one upward resisting force from the fluid pressure *p*. Since the weight increases with *h*, so does the fluid pressure, leading to the law of hydrostatic pressure. As a result, a body immersed in a liquid experiences a net upward force from the liquid, called buoyancy, of magnitude *B* equal to the weight of the displaced fluid (b).

the weight is small and therefore the fluid pressure is almost that of the external air pressure. In contrast, if we draw the surface near the bottom of the container, the fluid weight is large so the pressure must also increase a lot. As a result of gravity, the pressure in the liquid is therefore not constant but increases with depth.

These arguments of force balance can be used to calculate the spatial dependence of the pressure exactly. Let us denote by $p(h)$ the value of the pressure in the fluid at a depth h below the top surface and p_{top} the value of the pressure at the very top (equal to the air pressure). We use ρ to denote the mass density of the liquid, i.e. the mass of 1 cubic metre of fluid, and we assume that it is constant throughout the fluid. For a liquid such as water, we have $\rho_{water} \approx 1{,}000$ kg/m^3 while a gas such as air has a much smaller density, $\rho_{air} \approx 1.2$ kg/m^3.

The force balance on the total amount of fluid located between the top surface and the virtual surface located at a depth h below can be expressed mathematically. If the surface has total area A, the fluid volume is Ah and thus its mass is $m = \rho Ah$. The two downward forces acting on the fluid are therefore the weight, mg, and the force from the pressure at the very top, $p_{\text{top}} A$. To balance these forces is the upward resisting force due to the fluid pressure, of magnitude $p(h)A$. The equilibrium between these two forces means that $p(h)A = p_{\text{top}} A + \rho ghA$ and thus

$$p(h) = p_{\text{top}} + \rho gh. \quad (1)$$

This is the law of hydrostatic pressure showing that the pressure in the liquid increases linearly with depth.

This increase can lead to the generation of very large pressures. For example, scuba divers descending in the ocean experience 10,000 Pascals of added pressure for every metre of swimming below the surface. Human beings are known to be very sensitive to changes in pressure, and even casual swimmers diving to reach a toy at the bottom of a swimming pool experience the feeling of compression in their ears due to the increase in hydrostatic pressure. Equivalently, if pressure increases with depth, it also means that it decreases with height. For example, the pressure in the atmosphere decreases as one moves upward, explaining the need for aeroplanes to maintain cabin pressure during flight. Similarly, in plants, the fact that the pressure in the vascular system is approximately atmospheric near the plant roots and that it decreases with height from roots to stems means that the pressure can reach values significantly below atmospheric; for tall trees, that pressure can even become negative.

The law of hydrostatic pressure leads to the other remarkable property that the pressure only depends on the height h below the top surface, and has no horizontal dependence. The size of the container used to generate hydrostatic pressures is irrelevant, and if instead of a large bucket one uses a small tall tube, identical fluid pressures can be induced using much smaller amounts of liquids.

This is the principle exploited by an invention called the hydraulic press whereby hydrostatic pressures combined with large surfaces are used to generate large forces and lift very heavy objects. Legend has it that Pascal himself used the large hydrostatic pressure generated in a tall thin tube to burst open a large wooden barrel.

Buoyancy

The fact that the pressure in a fluid increases with depth leads to another important concept, that of buoyancy. This is illustrated in Figure 3(b). Take a rubber duck initially located outside a liquid container and proceed to completely submerge it in the fluid. To keep the toy immersed, one needs to apply a downward force to it. This is because when the rubber duck is inside the liquid it is subject to the same distribution of hydrostatic pressure as if it were not present in the fluid. The fluid pressure continues to increase with depth and therefore the pressure pushing the duck down is lower than that pushing it back up, resulting in a net force. This force due to the displaced fluid is called the buoyancy force, and its magnitude is exactly equal to the weight of the fluid displaced by the rubber duck. Indeed, if the duck was not in the fluid, the weight of the fluid would balance the pressure distribution on its surface. If you remove the fluid but replace it by the duck, the pressure distribution remains identical and thus it continues to apply the same force, equal to the opposite to the weight of the liquid.

This remarkable result is famously associated with Archimedes in ancient Greece. Long fascinated by geometry and mechanics, Archimedes was confronted with the problem of measuring the volume of an irregular shape (a crown). As he was taking a bath and saw the water rise as he entered the water, he discovered that the measurement could be done by immersing the crown in a liquid and measuring the volume of the displaced fluid. Legend has it that Archimedes then ran through the streets of Syracuse (Sicily) screaming 'Eureka', now the established motto for scientific discoveries.

Surface tension

Another important property of liquids at equilibrium is their cohesion. Consider a water droplet, for example a small raindrop attached to your window or your windshield. We all know that droplets tend to be round. In fact, if the droplet is small enough it will be exactly spherical (Figure 4(a)). If now this droplet is sandwiched between two surfaces, then external forces must be applied in order to squeeze the droplet into a thin pancake shape (Figure 4(a)). If these forces are relaxed, then the droplet will return to its original spherical shape. In other words, the droplet has the inherent property to want to remain round.

This cohesive property is called surface tension, and it has a molecular origin. Particles in the liquid interact with their neighbours through long-ranged attractive forces. It is therefore energetically favourable for a given particle to have as many neighbours as possible so as to decrease the overall internal

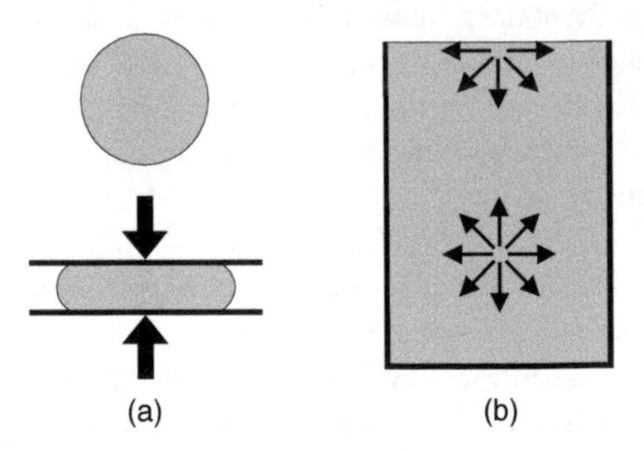

4. Liquid droplets have internal cohesion, so external forces are required in order to flatten them (a). This cohesion is quantified by a surface energy (surface tension) whose origin lies in the intermolecular forces in the liquid, which are stronger inside the liquid than on its surface (b).

molecular energy in the liquid. Particles in the middle of a liquid are surrounded by neighbours on all sides and have the minimum energy possible. In contrast, particles on the surface of a liquid only have neighbours on one side, the liquid side (Figure 4(b)). Compared to their friends deep into the liquid, surface particles are therefore missing out on half of potential molecular interactions. It is therefore energetically unfavourable for the fluid to have too many particles on the surface and the lowest energy state is obtained when the fluid has the smallest surface area possible so as to have as few surface particles as possible.

We therefore see that there is an energetic cost associated with increasing the surface area of the liquid. In order to increase the surface area one needs to provide external energy, and this is why external forces are required in order to squeeze the droplet in Figure 4(a). The surface tension of the liquid measures the energy of the liquid per unit of surface area. If surface tension were the only energy to consider, then drops would always be spherical, since a sphere is the geometrical shape with the minimum surface area for a given volume. This is famously the case in space, and astronauts have long entertained us with pictures of large floating balls of various beverages.

Wetting

The surface energy of a liquid arises from the intermolecular interactions between fluid particles. When a fluid drop is put in contact with a rigid surface an additional source of energy comes into play, namely that due to interactions between fluid and surface particles. Depending on the natures of both the liquid and the surface, fluid particles might favour interactions with other fluid particles over surface particles, or the opposite might be true and it might be energetically advantageous for the liquid to interact strongly with the surface instead of with itself. The branch of fluid mechanics studying the consequence of these interactions is called wetting.

Two broad categories of wetting behaviours can be distinguished. In the first type, called complete wetting, interactions with the surface are so favourable that the liquid spreads completely on the surface. This is a rather unusual case, but happens for example for water on ultra-clean glass. More common is the second situation, called partial wetting, where a small droplet deposited on the surface spreads until it takes the form of a portion of a sphere. The angle between the liquid surface and the rigid wall (measured inside the liquid) is called the contact angle. If that angle is acute, the surface is called hydrophilic and the droplet is less than half a sphere. This is the most common case in our everyday experience, happening for example for water on most metallic surfaces or treated glass such as windows or windshields (Figure 5(a)). If instead the contact angle is obtuse, the surface is said to be hydrophobic (so the drop is now more than half a sphere), with the most familiar example being water on plant leaves (Figure 5(b)).

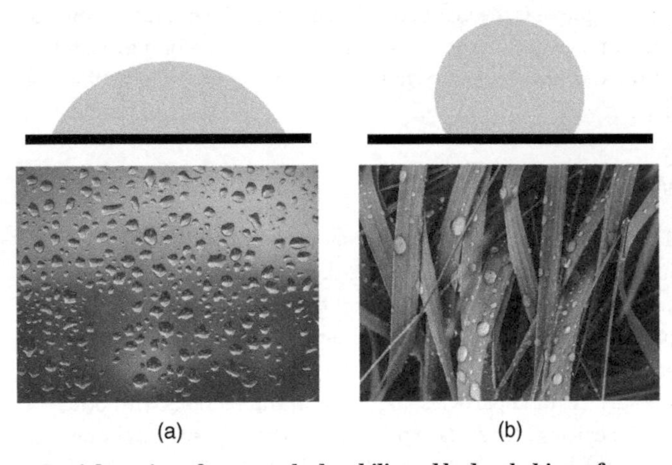

5. Partial wetting of water on hydrophilic and hydrophobic surfaces. In the hydrophilic case a liquid drop makes an acute angle with a surface, as seen for water on glass (a). In contrast, in the hydrophobic case the contact angle is obtuse, as illustrated in the case of water drops on grass (b).

The science of wetting is associated with the work of two illustrious scientists at the very beginning of the 19th century. Pierre-Simon Laplace was a French mathematician who derived an important equation showing that the presence of surface tension leads to an increase in pressure inside fluids with curved surfaces. He is now most famous for his work on the orbits of planets and for mathematical techniques; the Laplace transform is a classical method used to integrate differential equations. Not long after Laplace, the British physician and physicist Thomas Young worked on wetting and he derived a formula predicting the contact angle of a liquid as a function of the surface energies of the materials involved. Also well-known for his work on light, Young's name is famously associated with the mechanics of solids, and indeed the deformability of an elastic material is measured by its Young's modulus.

Chapter 2
Viscosity

We saw in Chapter 1 that the defining characteristic of fluids is their ability to flow in response to external forces. At its most fundamental level, the science of fluid dynamics consists in predicting how a specific fluid deforms in response to an applied forcing, and conversely what force is required to move a fluid in a prescribed manner.

Flow velocity

Fluid motion is quantified using the flow velocity. Since fluids can move in three independent directions of space, the velocity is a three-dimensional vector, denoted $\mathbf{u}$, with each component measuring the velocity in a particular direction. Furthermore, because the fluid velocity can take different values at different locations in space, it is called a field. For example, in a river the fluid on the riverbed does not move much while the flow is much faster near the water surface. The mathematical framework to study the behaviour of continuum fields is called field theory, with countless applications ranging from electromagnetism and gravitation to quantum physics.

In addition to its dependence on spatial location, the velocity in a fluid will often depend on time. In the river example, the flow might be slow in the autumn when temperatures drop but much stronger in the spring when mountain snow is melting. Traditionally we can quantify this time-dependence in two ways. The first one, called Lagrangian after the Italian-French 18th-century mathematician Joseph-Louis Lagrange, consists in considering a particular parcel of fluid and following it along its journey. When the fluid velocity increases, the parcel will have a positive acceleration, and when it decreases the acceleration will be negative. This manner of measuring time-dependence is intuitive to us as it is the way we experience acceleration in a car or on a roller coaster.

Unfortunately, the Lagrangian approach does not allow us to easily quantify and predict fluid motion as, in order to use it, we would need to be able to follow every parcel of fluid for all time. Instead, fluid mechanics has adopted a different approach termed Eulerian after the 18th-century Swiss Leonhard Euler, one of the greatest and most prolific mathematicians of all times at the origin of countless contributions in both pure and applied mathematics.

In the Eulerian approach, the time-dependence of the velocity is not measured by following fluid particles but instead by choosing a specific position in space, and measuring the changes of the velocity in time at that particular location. Using the example of the river, this is akin to choosing a spot on the river bank from which you stare at the river and keep track of the river velocity at that spot. Alternatively, it is like looking at cars passing by you on the side of the road and measuring how their speeds change in time. In the Eulerian approach, the velocity field is then denoted $\mathbf{u}(\mathbf{x}, t)$ where $\mathbf{x}$ refers to a fixed position in space and t is the time at which the velocity is measured.

Flow kinematics

Of course, fluids flow in many different ways, and the world around us abounds with examples. From waterfalls and rivers to boiling water and cigarette smoke, it looks as if the possibilities for fluid motion are infinite. However, despite this apparent complexity, it turns out that all flow kinematics, i.e. the mathematical description of how fluids can deform, can be described locally as the superposition of only three fundamental building blocks for fluid motion: translation, rotation, and extension.

These three flows are illustrated in Figure 6. The simplest possible flow is a uniform translation, illustrated in Figure 6(a) where, as is standard in the depiction of fluid motion, arrows are used to indicate the direction of flow and solid lines called streamlines show the motion of fluid particles. A uniform translation simply measures how fast the fluid moves locally on average. Notably, a rigid body could undergo a uniform translation as well, so no intrinsic deformation is associated with that flow.

The second fundamental flow is called a uniform rotation, as illustrated in Figure 6(b). In that case, the fluid is rotating around a fixed direction. Here again, this type of motion is not specific to fluids and a rigid body such as a ferris wheel could also rotate around a particular direction.

The third type of flow is called extension (or often pure extension to emphasize that it does not include any translation or rotation) and is illustrated in Figure 6(c). Qualitatively, in this flow the fluid

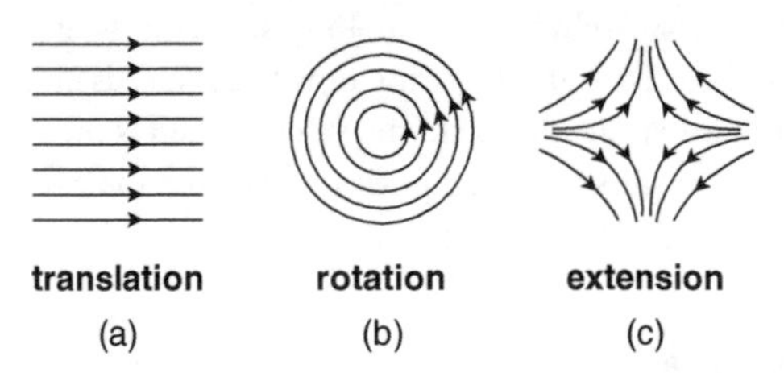

6. The three fundamental modes of fluid flow: uniform translation (a), solid-body rotation (b), and pure extension (c). While translation and rotation are not associated with deformation, a fluid has to deform when subject to an extension.

is contracting along one direction at a given rate (the horizontal direction in the figure) and extending along the other direction at the same rate (vertical direction). The streamlines are hyperbolas, which indicate that any fluid particle far from the centre of extension will be brought in (mostly) horizontally before being sent away (mostly) vertically. Importantly, while translation and rotation apply to both fluids and solids, a pure extension could not be experienced by a rigid body as it would have to be stretched and compressed to accommodate the shape of the streamlines. An extensional flow is therefore the most fundamental signature of local deformation in a fluid.

Incompressibility

In addition to the three fundamental flows described above, two additional flows can also be relevant. However, because they cannot be obtained as a superposition of translation, rotation of extension, they have to be considered separately. The first of these flows is called a source, and it corresponds to the net expansion of the fluid, similar to the type of motion that would occur if an elastic material was stretched equally from all directions (Figure 7(a)). The second flow is called a sink, and it is the inverse of a source (Figure 7(b)). In that case, the fluid is being

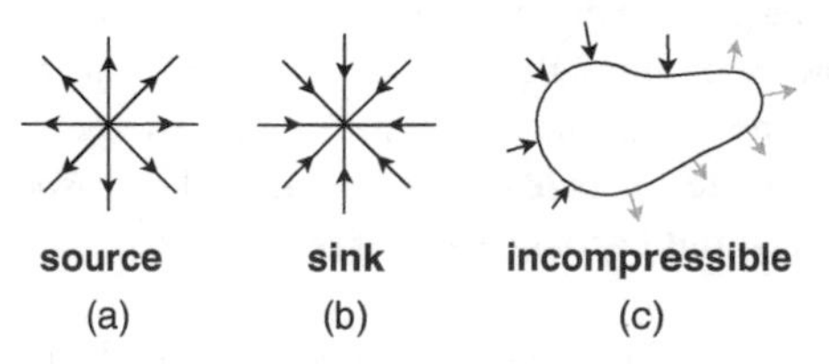

7. A source flow corresponds to the expansion of a fluid (a) while a sink leads to a net compression (b). Both types of flows are allowed only if the flow is compressible, meaning that a net amount of fluid can enter or leave a closed surface. If the flow is incompressible then neither sources nor sinks are allowed in the flow and the net flow through any closed surface is always zero (c).

compressed equally from all directions, and an intuitive analogy would be that of an elastic sponge being squeezed.

Sources and sinks are only allowed in flows if the fluid motion is compressible. A lot of research work has shown that this only happens if the velocities in the fluid are sufficiently large, at least on the order of the speed of sound in the fluid. The speed at which sound travels in water is approximately 1,500 m/s. For all water flows where the velocity is much less than this value, the fluid motion is essentially incompressible and sinks and sources cannot be part of the flow. Given the very large value of the speed of sound, this constraint is clearly not a severe one and all human experiences of water flow are in the incompressible limit. Exceptions will often occur in high-speed flows, for example for the flow around the rotor blades of submarines.

In the case of air, the speed of sound is roughly 300 m/s, and here also the flow of air with speeds well below this value is essentially incompressible. However, it is often counter-intuitive to think of the motion of a gas as being incompressible when gases can clearly be compressed, and indeed many of us have already compressed air when inflating a bike tyre. The subtle point concerns the difference between an incompressible fluid and an incompressible flow. As a gas, air is a fluid that can be compressed. But in order to compress the gas, significant pressures need to be applied to it, and for low-speed flows the generated pressures won't be large enough to induce any notable compression of the gas. Counter-examples include the air flow around high-speed jets able to break the sound barrier (see Chapter 4).

In this book we will only consider flows with speeds sufficiently small that they can safely be assumed to be incompressible. This is the relevant limit for all flows occurring in the living world, and in most industrial settings. An interesting consequence of incompressibility concerns the velocity field in the fluid, **u**. Indeed, if the flow is incompressible, its velocity cannot be completely arbitrary because it cannot contain any sources or sinks. More broadly, if we draw an arbitrary closed surface inside the fluid, the net rate of flow entering the surface has to be zero, as otherwise a net expansion or compression would be taking place (Figure 7(c)). So the velocity field must be such that all velocities through the surface somehow have to add up to zero, and that must also be true for any arbitrary surface drawn in the fluid. The mathematical consequence of this is that the velocity field is said to be divergence-free. The divergence of a velocity field computes its spatial variations in all three directions and when the divergence of the velocity is zero then the velocity varies by just the right amount to ensure that no volume of fluid can ever be compressed or expanded.

Fluid dynamics

In order to capture how a fluid deforms in response to forces, we need to carefully define what it means to apply a force on a moving fluid. In the absence of motion, we already discussed in Chapter 1 the two types of forces acting on fluid particles.

First, we discussed the force of gravity. This type of forcing is known as a volume force. If we consider a small volume V of fluid, each particle in the volume experiences the force of gravity (Figure 8(a)). Remove the fluid from the volume and the gravitational force disappears. This does not apply just to fluids,

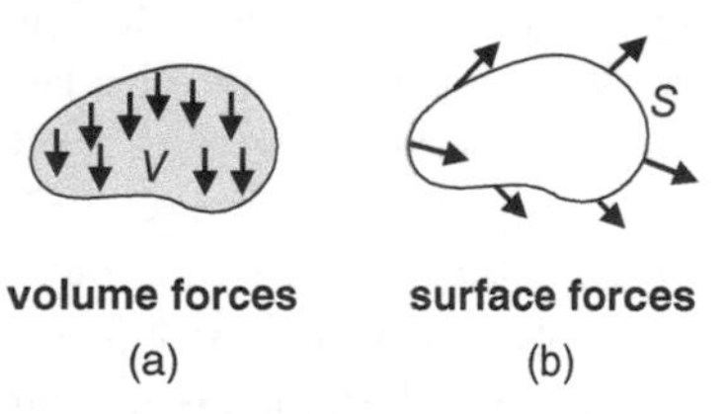

8. Volume force vs. surface forces. Gravity is an example of a volume force (a) acting on any fluid particle within a volume V. In contrast, surface forces (b), such as those due to viscous friction, are induced by the fluid motion immediately on the other side of the surface S.

and even for humans, gravity is a volume force: adults have larger weights than infants due to their larger mass. In most situations of interest, gravity is the only volume force to consider in a fluid, and it tends to play an important role for fluids on length scales of millimetres or more, while small-scale fluid mechanics is usually not affected appreciably by gravity. Another type of volume force is the electric force for fluids that carry net electric charges and are subject to an external electric field—a situation rarely encountered in our daily life but important in many industrial uses of ionic fluids.

The second type of force acting on a fluid discussed in Chapter 1 was the pressure. Recall that the pressure is the internal force pushing against a surface (per unit of surface area) and that the pressure is isotropic, meaning its value is the same in all directions. The pressure is an example of a surface force. In contrast to volume forces, surface forces act along, well, a surface. If we consider our small volume of fluid V bounded by a surface S with fluid on both sides, the surface forces are due to the fluid immediately on the other side of S acting on it, not due to the fluid particles inside the volume (Figure 8(b)). An intuitive experience of surface force in our daily life is the wind we feel when we ride a bike or stick a hand outside the car window: it is only due to the surface of our bodies (and not what is inside) that we experience this force.

Surface forces play a very important role in fluid motion as they often dominate the dynamics. As a result, scientists have given surface forces a special name, and they are referred to as tractions. The traction at a given point for a given small surface area S with fluid on either side of S is the force exerted by one fluid across the surface on the fluid located on the other side (per unit of surface area). The force due to the fluid pressure is an example of a traction, which acts directly perpendicular to the surface. Put simply, the pressure always pushes against the surface. When the fluid is not in motion, pressure is the only traction, and in that case it is called the static pressure.

One of the most fundamental questions in fluid mechanics is the intuitive understanding of what the surface tractions are when fluids are in motion. To help with intuition, let us consider a situation from our daily life. A heavy book sits on a table and you start pushing on it so as to make it slide across the table. The table experiences two surface tractions as a result of the presence of the book. First, the weight of the book acts as a downward force on it. Since the book is at equilibrium, the reaction force from the table balances the weight and therefore, by Newton's third law, the table experiences a downward traction perpendicular to its surface. Further, since the book is moving, it experiences a frictional force exerted by the table and therefore, due again to Newton's third law, the book exerts a frictional surface traction in the direction parallel to the tabletop. Two components of the traction are therefore acting on the table: one perpendicular to the surface of the table and one parallel to it.

The same decomposition into components applies for fluids in motion. If we consider a small surface drawn in the middle of a fluid in motion (Figure 9(a)), the traction acting on the surface has two components, one perpendicular to the surface and one parallel to it, and scientists use the word 'stress' to denote each component of the traction. The perpendicular traction is called the normal stress (Figure 9(b)) while the part of the traction acting parallel to the surface is called the tangential stress (Figure 9(c)).

In the absence of motion, there is no tangential stress and the only normal stress is the static pressure. How is this different when the fluid moves? Answering this question rigorously is difficult because it requires the use of complex mathematical machinery, but it can be intuitively summarized as follows. In addition to the static pressure present in the fluid at rest, the stress in a moving fluid is made up of two components. First, it is possible to separate

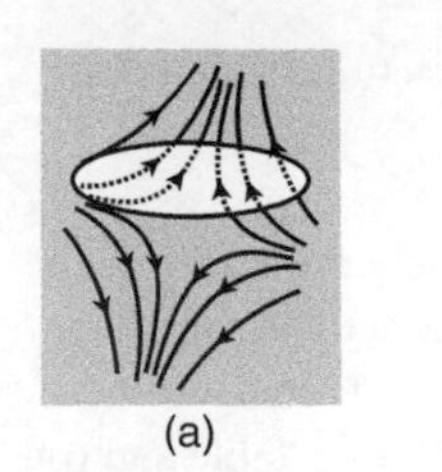

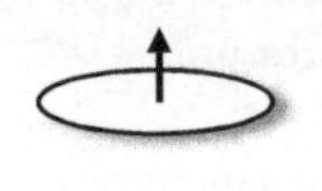

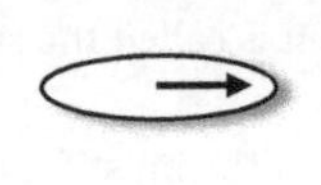

9. A small surface drawn anywhere in a fluid flow experiences, as a result of fluid motion, surface forces (per unit of surface area) called tractions (a). The component of the traction perpendicular to the surface is called the normal stress (b) while the component along the surface is the tangential stress (c).

mathematically the part of the normal stress that takes the same value for any orientation of the small surface, i.e. the isotropic part of the normal stress. This is called the dynamic pressure by analogy with the isotropic static pressure. When there is no flow, the dynamic pressure is zero. In contrast, in a fluid in motion the total pressure (i.e. the total component of the perpendicular force acting per unit of surface area that has the same value in all directions) is the sum of the static pressure and the dynamic pressure.

But the stress is not made up of just a pressure. The extra, non-pressure stress is due to friction in the fluid and is called a viscous stress. There is a normal viscous stress (so the total normal stress is the pressure plus the normal viscous stress) and there is a tangential viscous stress. Both viscous stresses are zero in the absence of motion. Physically, viscous stresses in a moving fluid are analogous to the frictional force experienced by the table in the example above as a result of the sliding motion of the book.

Because we have an intuitive understanding of the concept of static pressure, the dynamic pressure does not challenge our understanding. The pressure has one value when the fluid does not move and it is natural to expect that it would change under fluid motion. The genuinely new concept, and one of the most important notions in fluid mechanics, is that of a viscous stress. At its most basic level, viscous stresses measure all the non-pressure forces acting internally in the fluid due to its frictional resistance to motion. We know from our daily life that not all fluids flow with the same ease. It is straightforward for water to flow out of a bottle while it takes a lot of effort to squeeze honey out of one, and this difference is measured by an important material property of a fluid called the viscosity.

Viscosity

The simplest way to understand what the viscosity of a fluid represents is to use a simple thought experiment, often attributed to Newton and illustrated in Figure 10. Consider a simple fluid located between two parallel, identical surfaces of area A, which are separated by a distance h. The bottom surface is kept immobile while the top surface slides with steady speed U. Because of the friction in the fluid, an external force F has to be applied in order to make the surface move and when we no longer apply the force the motion eventually stops. The fundamental question is then: what is the value of F?

This classical experiment and multiple variations of it have been realized over the years—all coming to the same conclusion. First, with everything else being fixed, the force F is found experimentally to be linearly proportional to the surface area of the plates, A. Double the area, and you double the force. In other words, the important quantity is not so much the force as it is the

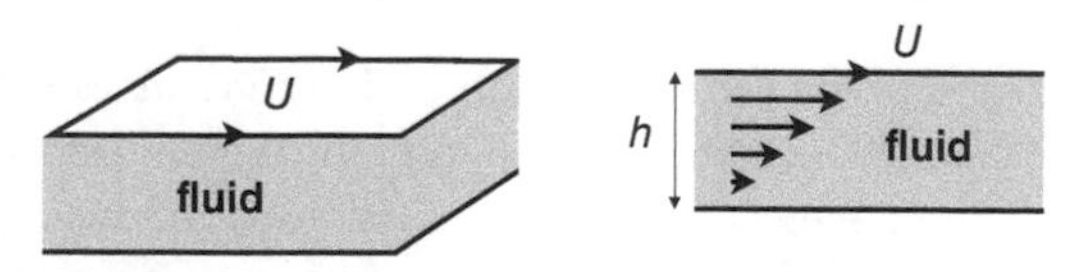

10. Experimental setup for Newton's empirical law of viscosity. A fluid is bounded by two rigid surface separated by a distance h. By measuring the force required to move one of the surfaces at speed U relative to the other we can define the dynamic viscosity of the fluid, μ.

force per unit of surface area, i.e. the tangential stress. Second, the force (and therefore the tangential stress) is found to vary linearly with the velocity of the top plate U. If you double the speed, you double the stress required to make the surface move at that speed. Third, the force is found to be inversely proportional to the distance h between the plates. If you divide h by two, the stress doubles and it is therefore harder to put fluids in motion in thin gaps than in wide geometries.

It is important to realize that these three properties of the dependence of F on the parameters of the experimental setup (linear variation with A and U and inversely proportional to h) could not have been postulated without carrying out an experimental study. They are not the consequence of a mathematical calculation or a grand theory but instead are empirical results capturing the observed behaviour of simple fluids in motion.

We can use however a mathematical formula to quantify Newton's observations. If repeated measurements are made for the values of F as a function of A, U, and h and we plot the ratio F/A as a function of U/h, we obtain a straight line passing through the origin, confirming that one quantity is directly proportional to the other. The ratio F/A is the tangential stress acting on the surface while the ratio U/h has dimension of inverse time and is called the shear rate. The linearity between tangential stress and shear rate for simple fluids can be written mathematically as

$$\frac{F}{A} = \mu \frac{U}{h}, \tag{2}$$

where the coefficient of proportionality μ is a constant called the dynamic viscosity—a material property of the fluid whose value depends only on temperature. The dimensions of the dynamic viscosity in SI units are given as kg/ms, which can also be written as Pa s, where we recall that a Pascal (Pa) is the unit used for pressure, i.e. 1 Pa = 1kg/ms^2. At room temperature, the dynamic viscosity of water is $\mu_{\text{water}} \approx 10^{-3}$ Pa s, while the value for air is smaller, $\mu_{\text{air}} \approx 1.8 \times 10^{-5}$ Pa s.

The mathematical law in Equation (2) is now known as Newton's empirical law of viscosity. It agrees with our physical interpretation of the viscosity as an internal fluid friction: a more viscous fluid experiences more friction and therefore requires a larger external force to be applied in order to be moved at a given speed. Furthermore, the relationship in Equation (2) can be used to clarify the classification of simple vs. complex fluids first proposed in Chapter 1. Simple fluids are those that always obey Newton's linear empirical law of viscosity and they are therefore called Newtonian fluids. Complex fluids then refer to any material with non-Newtonian properties.

Shear flow

The setup shown in Figure 10 is called a shear experiment. While it is apparently very simple, it has far-reaching consequences. By measuring the force on the moving plates one can directly calculate the viscosity of the fluid, and therefore shear flows are used in viscosity-measuring devices called rheometers. The field of rheology is devoted to characterizing the material properties of fluids and all rheometers use versions of this shear experiment in order to measure viscosities.

Beyond viscosity measurement, the shear experiment allows us to learn more about fluid motion. Using small tracer particles one can characterize the flow field directly between the two surfaces. By doing so, we learn two important facts about moving fluids. The first one is that the velocity of the fluid close to any surface is equal to the velocity of the surface. So the fluid near the bottom stationary surface does not move while the fluid near the moving plate moves with velocity U. This property is a fundamental fact of all fluid motion and is called the no-slip boundary condition. Intuitively, it is well known that if you dip your toes in a pond or your hand in a bath, they will still be wet after you remove them. Indeed, if a surface is put in contact with a liquid and then removed, the no-slip condition implies that the nearby fluid cannot help but move along with it. So you can start blaming no-slip for getting wet.

A second important fact about the flow in Figure 10 is the variations of the velocity across the experimental setup. We saw that the flow speed was equal to zero at the bottom and equal to U at the top. As it turns out, the flow speed varies exactly linearly between these two values across the fluid. If y denotes the coordinate perpendicular to the surfaces, equal to 0 at the bottom and h at the top, the fluid speed at position y is then obtained as $u(y) = Uy/h$. This linear flow is called a shear flow. Since the flow is linear, the rate of change of the velocity across the fluid (i.e. how much u changes as y changes) is constant, and equal to the shear rate, U/h. A shear flow is also known as a Couette flow, named after the French engineer and scientist Maurice Couette, who at the end of the 19th century carried out a lot of fundamental work on the no-slip condition and viscosity measurements using simple flow setups.

A final property of fluid motion highlighted by a shear flow is that of reversibility. The result for the force in Equation (2) shows that if the plate speed is changed from U to $-U$, the force F is transformed into $-F$, i.e. the force remains of the same magnitude but it acts in the opposite direction. This is called the reversibility property of viscous flows, and it implies that in order to apply a net force on a viscous fluid, in general a simple back-and-forth action is not very efficient. Indeed, in the shear experiment if you move the plate with velocity U for a period of time and then reverse the motion and move with speed $-U$ for the same amount of time, the total force applied on the fluid is zero. Small biological appendages called cilia used by cells to transport fluid, for example for mucus clearance in our lungs, deform periodically using a back-and-forth motion and therefore have to find a way to bypass this reversibility constraint. They do so by undergoing motion that is asymmetric, with large deformation in one half of their beating stroke and a smaller one in their return stroke.

The Navier–Stokes equations

In this chapter, we highlighted a number of important properties of fluid motion, from flow kinematics and deformation to stresses and viscosity. These ideas were put on a strong mathematical footing in the 19th century by two scientists whose names are forever associated with fluid mechanics: Claude Navier, from France, and George Stokes, who was born in Ireland and carried out his work in Cambridge. We now call the set of mathematical equations describing the dynamics of the velocity ($\mathbf{u}$) and pressure fields (p) in an incompressible Newtonian fluid the Navier–Stokes equations. Although we will not attempt to derive or solve them, it is instructive to reproduce them here, if only to admire the beauty of their mathematical structure. Using t to denote time and ρ the mass density of the fluid, the Navier–Stokes equations are written formally as

$$\nabla \cdot \mathbf{u} = 0, \quad \rho \left(\frac{\partial \mathbf{u}}{\partial t} + \mathbf{u} \cdot \nabla \mathbf{u} \right) = -\nabla p + \mu \nabla^2 \mathbf{u} + \rho \mathbf{g}, \qquad (3)$$

where the gradient ∇ is a differential operator that involves spatial derivatives of the fields and ∂ is the symbol for partial differentiation. Many mathematical subtleties and challenges remain associated with these equations (see Chapter 8).

Chapter 3
Pipes

The shear flow discussed in Chapter 2 allowed us to gain intuition on the concept of viscosity and viscous shear stresses. The most common flow occurring in industry and in nature is, however, slightly more complex than a linear shear flow, and consists in a fluid pushed through a rigid pipe by an excess pressure. A careful study of elementary pipe flow turns out to reveal a lot on the fundamental physical behaviour of fluids in motion. In this chapter, we will explore the flow of fluids in pipes, its importance in nature and applications in industry, and its historical role in our understanding of turbulence.

Pipe flow

Everything we need to know about pipe flows can be drawn from the elementary case of a straight pipe with a circular cross section. This prototypical pipe flow is illustrated in Figure 11(a). The pipe is rigid and filled with a viscous fluid. The fluid at the pipe outlet is at atmospheric pressure (here denoted by p_{atm}), while the pressure in the fluid at the inlet is imposed to be $p_{\text{atm}} + \Delta p$. A difference in pressure, Δp, is therefore applied between the inlet and the outlet.

When $\Delta p > 0$, this extra pressure drives a fluid flow from left to right in Figure 11(a). This situation occurs, for example, when one blows air bubbles in a glass of water using a straw or when our

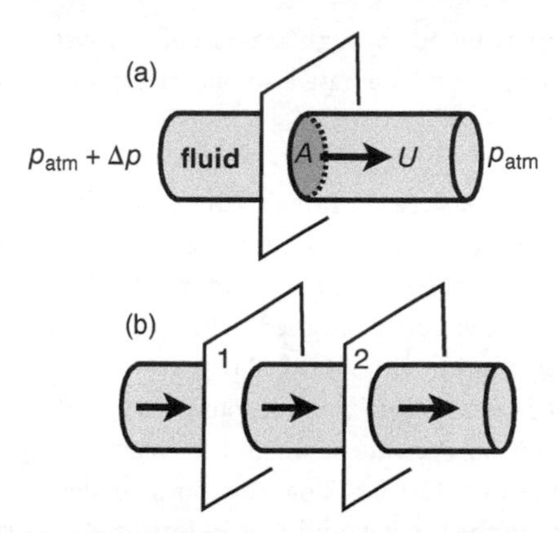

11. A difference in pressure (Δp) is imposed between the ends of a rigid pipe, which results in fluid flow (a). When the pressure is higher on the left ($\Delta p > 0$), flow occurs from left to right. The flow rate in a pipe is the product of the cross-sectional area of the pipe, A, and the mean fluid speed, U. In an incompressible flow, the flow rate is the same through any cross section of the same pipe, as otherwise the fluid would be compressed or extended (b).

beating hearts push blood through our arteries and veins. Another option is possible, namely one where the difference in pressure is imposed to be negative, i.e. $\Delta p < 0$. In that case, the flow direction is reversed and fluid moves from right to left in the figure. This happens when we use a straw to suck a drink from a soda can; a similar situation arises in vascular plants when water flow travels upwards drawn by the negative pressures in the leaves.

Flow rate

A fundamental question about pipe flow is the determination of the rate of flow in the pipe as a function of the applied pressure difference. Clearly it is easier to suck water through a straw than honey, so the viscosity of the fluid is a key parameter in determining the flow speed. It is also known that partially blocked arteries and veins create tremendous challenge to our beating hearts pumping blood through the circulatory system, so the cross-sectional size of the pipe also plays a key role in setting the speed of flow.

How does one measure the rate of fluid flow? The intuitive way to think about this is to picture an open pipe with a liquid flowing out of it, and placing an empty bucket at the exit point. At the beginning of the experiment, the bucket is empty and then it progressively fills up with liquid. The natural way to measure the typical speed of the liquid is to measure the rate at which the bucket fills up, i.e. the rate at which the volume of liquid in the bucket increases with time. That rate does not depend on the size or the shape of the bucket, and it is an intrinsic property of the fluid and of the pipe through which it is being transported.

The rate of fluid flow, conventionally called the flow rate and denoted by Q, measures the volume of fluid going through the pipe per unit of time. If the velocity of the fluid in the pipe is known, the flow rate may be computed as illustrated in Figure 11(a). Imagine drawing an invisible surface crossing the pipe and perpendicular to the flow direction (therefore the fluid goes continuously through the surface). We call A the cross-sectional area of the surface in the pipe. To measure the flow rate, one must first calculate the mean value of the fluid velocity through the surface, denoted by U. During a time t, the fluid moves through the surface of area A by a distance $U \times t$ and therefore the volume of fluid going through the surface is $A \times U \times t$. Since the flow rate is the volume of fluid per unit of time, we obtain the classical formula $Q = AU$.

For incompressible flows, an important property of the flow rate is that, at any given moment, it takes the same value everywhere along the pipe. Indeed, let us imagine a pipe flow in which two different values of the flow rate exist, say Q_1 at some upstream location 1 and Q_2 downstream at position 2 (both locations are represented by the invisible surfaces drawn in Figure 11(b)). If $Q_1 > Q_2$ then more fluid enters the region between 1 and 2 than leaves it, so the fluid is accumulating there and is therefore being compressed. Similarly, when $Q_1 < Q_2$ more fluid is leaving the region between 1 and 2 than flowing into it, and thus the fluid is expanding. For an incompressible fluid, the only possibility is therefore $Q_1 = Q_2$. Since locations 1 and 2 could be anywhere along the pipe, we get the result that the flow rate has the same value everywhere in the pipe.

Note, however, that the value of the flow rate is allowed to change with time. For example, the flow out of the valve of a bike tyre is fast when the tyre is fully pumped up and decreases with time as the tyre gets deflated. But at any instant the flow rate at every point along the valve has the same magnitude.

First pipes

Since pipes play a crucial role in both biology and industrial applications, it is only fitting that pipe flow was simultaneously discovered by a doctor and by an engineer.

Jean Poiseuille was a 19th-century French doctor and a scientist. Believed to have been the first person to use a mercury manometer to measure blood pressure, he was also interested in understanding the physical principles behind blood flow. In a series of landmark studies in the 1840s, he investigated empirically the flow of human blood in narrow glass tubes. He discovered that the rate of flow through the pipe increased linearly with the applied pressure difference and with the fourth power of the pipe diameter. The Poise, a unit of viscosity corresponding to 0.1 Pa s, is named after him.

Around the same time as Poiseuille's work, but independently, the German engineer Gotthilf Hagen studied the flow of water in thin brass tubes. Hagen discovered the same empirical dependence of the rate of flow in the pipe on its diameter as Poiseuille. Hagen is often viewed as the father of hydraulics, the field of study of the generation and transmission of power using pressurized liquids.

Hagen–Poiseuille law

The law discovered by Poiseuille and Hagen now bears both their names. Considering a pipe of circular cross section with diameter D and total length L (see Figure 12), the Hagen–Poiseuille law states that the flow rate Q in the pipe of a fluid of viscosity μ is proportional to the applied pressure gradient as

$$Q = K\frac{D^4}{\mu L}\Delta p, \tag{4}$$

where K is a universal constant for all fluids. The proportionality of Q to Δp and D^4 and to the inverse of the length of the pipe, $1/L$, were all obtained empirically by Poiseuille and Hagen. The inverse dependence on the viscosity of the fluid tends to also be attributed to them but this was in fact derived theoretically more than twenty years later.

We can interpret the result in Equation (4) in two ways. Consider a fluid in a given pipe. The standard interpretation of Equation (4) is to give the value of the fluid flow rate induced by a given

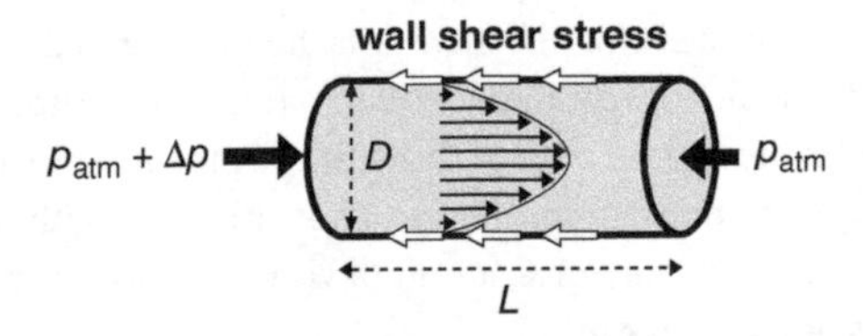

12. The fluid velocity in a pressure-driven pipe flow varies parabolically across the pipe, with no fluid motion on the wall and maximum speed along the pipe axis (here the pipe is cylindrical with diameter D and length L). The equilibrium of forces on the fluid means that the sum of all wall shear stresses distributed on the surface of the pipe (white arrows) has to be equal to the total force due to the fluid pressure (black arrows).

pressure difference. However, one could also see the result as an equation for the pressure drop Δp that would be required in order to create a specific flow rate Q through the pipe. Both interpretations are equally valid, and the relevant one depends on the specific application of interest. For example, many industrial systems generate high pressure by relying on hydrostatics (see Chapter 1), a situation where the excess pressure would therefore be fixed but the resulting flow rate unknown. In contrast, in a delivery system for drinking water, the design criterion is the flow rate accessible to the inhabitants at the tap, so in that case the flow rate would be known but the required pressure unknown.

Pressure-driven pipe flow

Among his many achievements, Stokes was the first to derive a theory predicting exactly the results captured by the Hagen–Poiseuille law for viscous flows (although he did not publish his theory at the time . . . because he did not believe it). Stokes mathematically solved for Newton's law at every point inside the fluid (i.e. the Navier–Stokes equations), and obtained a number of remarkable predictions, all of which turned out to be correct.

First, Stokes obtained the result that the pressure in the fluid decreases linearly from the inlet to the outlet. For example, if the pressure at the entrance is $p_{\text{atm}} + \Delta p$ and at the outlet p_{atm}, the pressure halfway through the pipe is $p_{\text{atm}} + \Delta p/2$.

Second, Stokes showed that the flow is directed everywhere along the direction of the pipe. In other words, at every location inside the moving fluid, the local fluid velocity is a vector whose direction is parallel to the axis of the pipe, and no flows are swirling around in the other two dimensions. Flows with this property are called unidirectional flows.

Third, Stokes demonstrated that the magnitude of the velocity in the fluid varies quadratically across the pipe. The minimum value of zero velocity is obtained on the pipe itself, as a result of the no-slip boundary condition, while the maximum flow speed occurs along the axis of the pipe (see illustration in Figure 12). In between, the flow magnitude is parabolic.

Finally, the remarkable property that the constant K in Equation (4) is a universal constant valid for all fluids raises an obvious question: What is its value? That constant was computed by a number of authors independently from one another, approximately 15 years after Stokes's original calculation, and the exact result is $K = \pi/128 \approx 0.025$.

What are the typical orders of magnitude for pressures in everyday life? Let us consider the simple case of sipping water ($\mu_{\text{water}} = 10^{-3}$ Pa s) through a straw of diameter $D = 1$ cm and length $L = 10$ cm. A child who wants to drink water at a leisurely rate of $Q = 10$ cm ^{3}/s using the straw would need to sip by lowering the pressure in her mouth (compared to the atmospheric pressure) by a value obtained from the Hagen–Poiseuille law (Equation (4)) as $\Delta p \approx 4$ Pa. This difference in pressure is quite small, similar in magnitude to the pressure applied by a mass of 40 mg spread over a surface of 1 cm^2.

Scaling vs. exact results

One of the beauties of mathematical results such as the flow rate in Equation (4) is that, in many instances, the exact value of the constant K does not really matter. This might seem like a bizarre statement to make. After all, if somebody is designing a piping system that needs to deliver a given amount of fluid, they need to know how to design the pressure differences along their pipes, and thus the value of K. And indeed, for any practical applications of fluid mechanics, it is crucial to have complete knowledge of all the input–output relationships for the fluid.

A first important point to note is that the value of K does not need to be derived mathematically as it can be obtained empirically. Since K is a universal constant, one can determine its value using a single experiment. Specifically, pick a pipe of dimensions D and L and a fluid of viscosity μ, apply a pressure difference Δp and measure the flow rate Q (or vice versa). The ratio between the terms on the left of Equation (4) and those on the right then gives access to the value of K. Since K is a universal constant, it has the same value for all fluids and in all pipes.

Second, the result in Equation (4) contains a lot of information on the fundamental behaviour of fluids in motion—even with the value of K unknown. For example, with everything else being fixed, Equation (4) says that the flow rate is linear in the pressure drop, so in order to create a flow twice as fast one needs to apply twice as much pressure. This type of result is widely used in fluid dynamics, and is called a scaling relationship. We write it as $Q \propto \Delta p$, to emphasize the linear relationship between two specific quantities (here the flow rate and the drop in pressure) without changing any of the other parameters.

The result in Equation (4) contains multiple scaling relationships. The most dramatic one is the scaling for the flow rate, $Q \propto D^4$, predicting a very steep dependence of the flow rate on the pipe diameter. If you divide D by 2, the flow rate will be divided by 16. If the pipe has a cross-sectional shape different from a circle, of course the exact result in Equation (4) is modified but the scaling with the typical cross-sectional length scale remains true. We also have the relationship $Q \propto 1/\mu$ so a fluid twice as viscous will flow half as fast. Similarly, the scaling $Q \propto 1/L$ indicates that very long pipes experience very small flow rates.

Alternatively, if Equation (4) is seen as an equation not for the flow rate for a given pressure but instead for the excess pressure required to drive flow at a given rate, we obtain the important scaling $\Delta p \propto L$. In a piping system to distribute liquids at a given rate, the applied pressure scales with the length of the pipe. We also see that $\Delta p \propto \mu$ and therefore, for example, in order to push honey, which is about 1,000 times more viscous than water, through a straw, the required pressure is 1,000 larger than that necessary to drive water at the same flow rate—the use of a straw is therefore not recommended in this case. Finally, we see that $\Delta p \propto 1/D^4$ so the pressures required to drive fluids in small pipes can quickly become very large.

Wall shear stresses

As the fluid moves along in a pipe, one of its important features is the fact that it exerts a force on the walls of the pipe. Similarly to Newton's experiment where an external force had to be applied to the top surface in Figure 10 in order to make the fluid move, here an external force would also need to be applied to the pipe in order to hold it in place. If not, the fluid will entrain the pipe.

Since the force applied by the fluid is distributed all over the surface of the pipe, the natural quantity of interest is the force per unit of surface area, i.e. the traction. Because the flow of fluid is parallel to the wall of the pipe, we are interested in the force directed with the flow, which we saw in Chapter 2 is called the tangential stress. The value of a tangential stress on a rigid surface is usually referred to as the wall shear stress and denoted by σ.

One way to calculate the value of σ is to exploit the exact solution for the flow derived by Stokes, and then use it to deduce the shear rates in the fluid. With these rates, we can then use Newton's postulate, Equation (2), to relate stresses in the fluid to the shear rates, and then to calculate the value of the wall shear stress. This classical result requires the use of calculus, and is therefore beyond the scope of this book.

Alternatively, we may compute the value of σ by bypassing calculus altogether and instead making use of Newton's third law. If σ is the force (per unit area) exerted by the fluid on the pipe then a force of the same magnitude σ is exerted by the pipe on the fluid per unit of surface area (in the opposite direction). The fluid in the pipe moves at constant speed and does not accelerate or decelerate so the entire fluid in the pipe must not experience any net force. Remember that Newton's second law says that, in order for bodies to accelerate or decelerate, a net force must be applied to them. So no acceleration means zero force.

What are the forces applied to the entire fluid in the pipe? There are only three of them, as illustrated in Figure 12. First the entrance and exit of the pipe are subject to the external pressures. As shown in the figure, the pressure on the left is $(p_{\text{atm}} + \Delta p)$ acting on the cross-sectional area of the pipe, $\pi D^2/4$ and contributing therefore a total force of $(p_{\text{atm}} + \Delta p)\pi D^2/4$ pushing the fluid in the direction of the flow. Similarly, the pressure p_{atm} at the exit of the pipe leads to a total force of magnitude $p_{\text{atm}}\pi D^2/4$ opposing the flow. The net force from the pressure difference between the inlet and outlet forces is therefore given by $\Delta p \pi D^2/4$ in the direction of the flow.

Opposing this force is the resistance from the pipe. Since the force per area offered by the pipe on the fluid has the magnitude of the wall shear stress, σ, and the total area of a pipe of length L is πDL, the total frictional resistance from the pipe has magnitude $\pi DL\sigma$. Since all forces on the fluid need to balance, we have the equality $\Delta p \pi D^2/4 = \pi DL\sigma$ and therefore the magnitude of the wall shear stress is obtained as

$$\sigma = \frac{1}{4}\frac{\Delta p D}{L}. \tag{5}$$

Remarkably, the value of σ does not rely on the knowledge of the flow speed in the pipe, and it is instead a necessary consequence of

Newton's second law. Furthermore, for flows driven by a given pressure drop Δp, the wall shear stress σ does not depend on the value of the viscosity of the fluid. How is it possible that water and honey, which have viscosities with three orders of magnitude difference, would lead to the exact same shear stress? The key to understanding the intuition behind this result lies in the scaling results obtained above. When the viscosity of the fluid increases, the Hagen–Poiseuille law, Equation (4), says that the flow rate increases as $Q \propto 1/\mu$. Since the flow rate is $Q \propto U D^2$ and the size of the pipe is fixed, this means that the average speed in the pipe varies also like the inverse of the viscosity, $U \propto 1/\mu$. Newton's law of viscosity, Equation (2), predicts that the shear stress in the fluid is the product of the shear rate and the viscosity. Since the shear rate is proportional to the velocity in the pipe, and since the velocity is inversely proportional to the viscosity, the shear stress in the pipe remains constant. Physically, when a fluid of different viscosity is used, the magnitude of the fluid velocity in the pipe adjusts so that the product viscosity $\times$ velocity remains constant.

If, instead, the flow in the pipe were controlled by fixing its flow rate, the value of the wall shear stress would scale very differently. Using the Hagen–Poiseuille law from Equation (4) we may rewrite Equation (5) as

$$\sigma = \frac{1}{4K}\frac{\mu Q}{D^3}. \tag{6}$$

The wall shear stress continues to vary linearly with the forcing in the problem (here the flow rate) but now it increases with the viscosity of the fluid, $\sigma \propto \mu$. Pushing a fluid through a pipe with a given flow rate is akin to setting the speed of the fluid everywhere in the pipe, and therefore the shear rates in the fluid flow. Following from Newton's law of viscosity, we thus see that the shear stresses are proportional to the fluid viscosity. We also note in this case that $\sigma \propto 1/D^3$ and small pipes can be subject to very large stresses.

Biological pipes

Arguably, the most famous example of flow in pipes is the blood circulation of animals. Blood is a watery suspension of cells and other components (in particular platelets) that behaves on all but the smallest length scales like a simple viscous fluid with a viscosity of about three times that of water. The purpose of the blood circulation is to bring red blood cells carrying carbon dioxide to the lungs where gas exchange takes place, wherein carbon dioxide diffuses out of the cells while oxygen diffuses in. The red blood cells then carry the oxygen throughout our body and deliver it to all cells and tissues, replacing it by carbon dioxide, a waste product of chemical reactions. Red blood cells then return to the lungs and the cycle continues.

The pump in the blood circulation is, of course, the heart, which contracts periodically, sending pulsatile flow through the incredibly intricate network of arteries (oxygenated blood from lungs to tissues) and veins (deoxygenated blood back to the lungs). Early enquiries on haemodynamics, the fluid dynamics of blood and the circulatory system, consisted in attempting to measure the geometry of the pipe network (see Figure 13(a)). Advances in anatomy means we now have a good understanding of the geometry of circulatory system from the largest arteries (a few centimetres in diameter) to large capillaries (a few hundred micrometres), as shown in all modern textbooks (Figure 13(b)). To illustrate the complexity of the network of vessels at even smaller length scales, we reproduce in Figure 13(c) observations of all blood capillaries in a 1-cubic-millimetre portion below the surface of the brain cortex of a mouse, as observed using fluorescence microscopy. This striking image conveys the complexity of the blood delivery system at the cellular level, with a high-density network of curved, connected pipes with a wide range of shapes and sizes.

Wall shear stresses play a particularly important role in the blood circulation. Termed in that case vascular shear stresses, they are

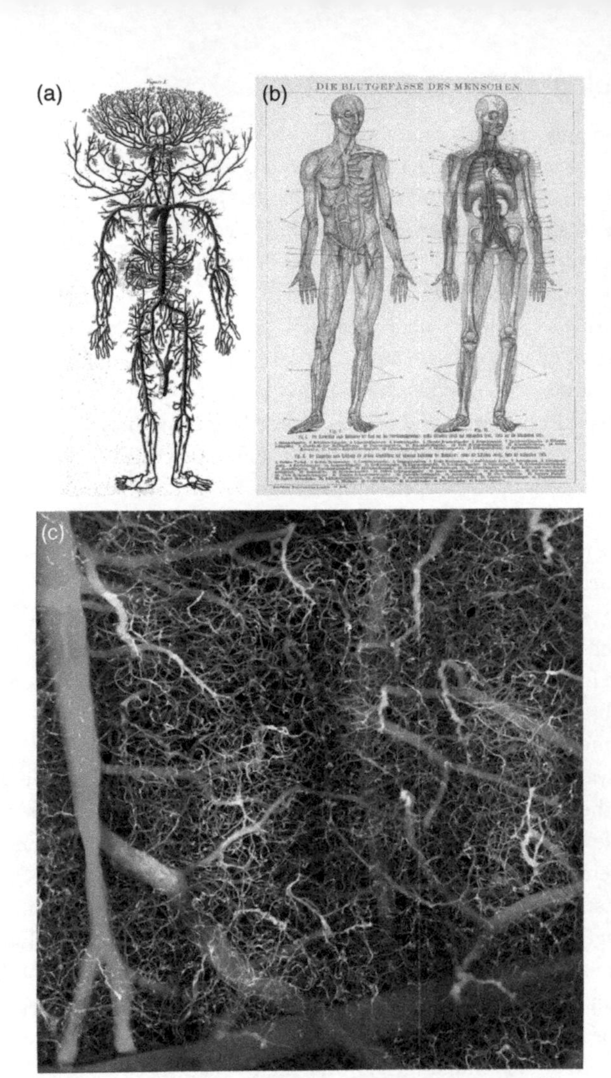

13. The anatomy of blood vessels as was known in the 18th century (a) is very different from the current textbook illustration of the circulatory system showing the large arteries and veins (b). Advanced imaging techniques can be used to get detailed information on blood vessels, here the blood capillaries in a 1-cubic-millimetre portion of a mouse brain cortex using fluorescence microscopy (c).

exerted on the endothelium, a layer of cells that lines the interior of blood vessels. Both the blood pressure, which acts perpendicularly to walls of the blood vessels, and the vascular shear stresses, which exert forces parallel to the vessels, apply time-varying mechanical loading and regulate numerous biological processes, including the thickness of blood vessels and the release of chemicals. Under abnormal conditions, vascular shear stresses can lead to remodelling of the circulatory network and to the creation of new blood vessels. Fluid mechanics therefore plays a crucial role in many pathological conditions such as heart disease and cancer.

For example, we saw above that pipe flows driven by fixed pressure have wall shear stresses that increase very quickly as $\sigma \propto 1/D^3$ with a decrease in the pipe diameter. This scaling result explains why arteries blocked by plaque (fatty deposit resulting from unhealthy diet) are so dangerous. Indeed, plaque leads to a decrease of the effective diameter D of the vessel, and therefore an increase in the wall shear stresses σ. This not only leads to damage of the biological tissues but can also lead to detachment of the plaque from the blood vessels, which may then travel to another blood vessel and block it, resulting in a heart attack or a stroke.

Pipes networks and hydraulic circuits

The blood circulation is an example of a pressure-driven flow inside a network of pipes. Oxygenated blood leaves the heart via the aorta, which then splits into multiple smaller arteries, themselves branching out into smaller and smaller blood vessels down to the smallest capillaries. The drinking water system in a house is organized in a very similar manner. There is a single entry point linking the house to the city water service. The main water pipe is then split at different points in a geometrical manner that allows water to be brought to all taps, showers, and toilets. From a fluid dynamics point of view, a pipe network can be illustrated as in Figure 14(a) where an inlet flow is split into many different

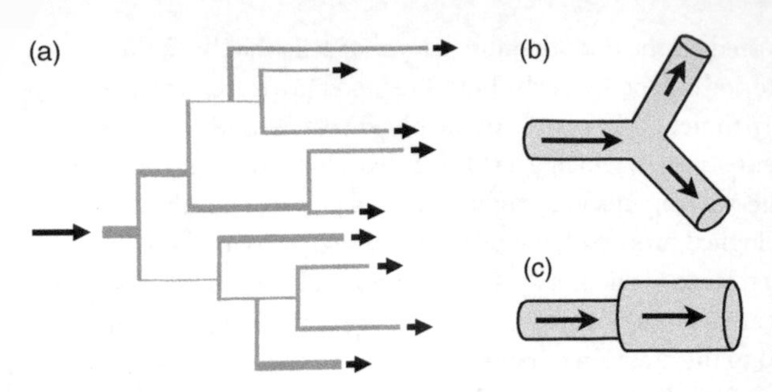

14. In a network of pipes, fluid enters at the inlet of the network and can exit at any of the outlets, with exit flow rates that depend on the overall geometry of the network (a). In the pipe network, composed of branches, the fluid flow divides in such a way that the input flow rate is equal to the sum of the output flow rates (b), so when an input pipe changes into a single output pipe with a different size, the flow rate remains constant (c).

pipes, all of different lengths and diameters, and flow is delivered at multiple outlets.

The two fundamental elements that make up a network are a junction in which one pipe (the mother pipe) splits into two (daughter) pipes, and one where an input pipe changes into a single output pipe of different size; both cases are illustrated in Figures 14(b) and 14(c). In order to determine how the flow from the upstream pipe affects the downstream pipes, we need equations relating the inlet and outlet pressures to the flow rates in all pipes. The intuitive method to solve this consists in rewriting the Hagen–Poiseuille law for any pipe, Equation (4), as $\Delta p = RQ$ where R is called the hydraulic resistance. At a junction, since the fluid is incompressible, the total input flow rate in the mother pipe has to be equal to the sum of the flow rates in the daughter pipes (Figure 14(b)). In contrast, when a single pipe changes into one of different size, the flow rate remains constant and instead it is the pressure differences along the two pipes that are added up to obtain the global pressure difference (Figure 14(c)).

Remarkably, these laws are the same mathematical equations one would have to solve in an electric circuit where various resistors are organized in series and parallel. Here, the flow rate plays the role of the electric current while the difference in pressure between two points in the pipe network is analogous to the difference in voltage between different points in the electric circuit. The conservation of flow at a junction is the fluid analogy to Kirchhoff's current law, stating that at any junction the sum of currents flowing into the junction equals the sum of currents flowing out of the junction. Since the equations are the same, all classical results on the effective resistance of electric circuits can be adapted to the case of pipe networks. For example, the hydraulic resistance of two fluid pipes in series is the sum of the resistances while, when two pipes are organized in parallel, it is the inverses of the resistances that are added up.

Laminar vs. turbulent flow

Because of their very small length scales, flows in small pipes such as blood capillaries always follow the Hagen–Poiseuille law. The fluid flow is always smooth and has all the characteristics described earlier. The flow in that case is called laminar. A fascinating phenomenon arises when one increases the magnitude of the fluid velocity in the pipe. In an experiment, this could be done by increasing the diameter of the pipe and the applied pressure, and using a fluid with a low viscosity such as water. When the velocity becomes too large, it turns out that the fluid flow changes dramatically in nature, which is the basis for one of the most important discoveries in all of fluid dynamics: the transition to turbulence.

Osborne Reynolds was an Irish-born British engineer and professor at what is now the University of Manchester. Among his many scientific contributions, he is best remembered for an 1883 landmark paper in which he described what happened to the flow of water driven in various pipes by gravity, using coloured water to

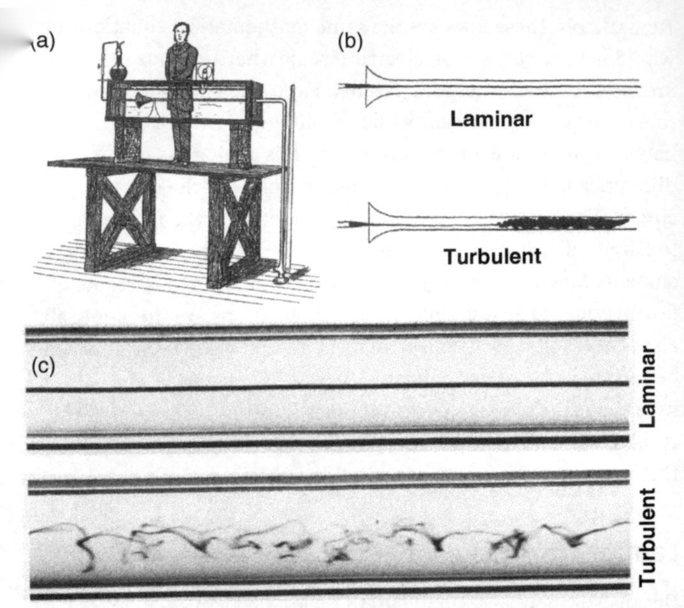

15. Reynolds's 1883 experiment was the first to report the transition from a laminar fluid flow to a turbulent one. The sketches from Reynolds's original paper show him standing next to his pipe-flow apparatus (a) and his observations of laminar and turbulent fluid motion (b). Experimental pictures illustrate the pipe flow of dyed water in both the laminar and turbulent states (c).

visualize the fluid velocity. The sketch illustrating his experimental setup is reproduced in Figure 15(a). His observations would change the field forever.

What Reynolds first reported was that the flow in the pipe was laminar at small speeds, in agreement with the earlier discussion. So far, no surprise. In fact, if he had measured the flow rate he would have obtained a nice agreement with the Hagen–Poiseuille law. However, upon an increase in the fluid speed, he observed that the flow took a completely different form. Instead of being smooth and unidirectional, the flow was full of sinuous eddies that

flowed in 'the manner so familiar with smoke'. In his paper he sketched his observations, which we reproduce in Figure 15(b) along with experimental pictures showing a similar result using dyed water in Figure 15(c).

What Reynolds observed is now called turbulence. Flows at low speeds are always laminar, steady, and well-behaved. Even in complex geometries it is fairly easy to predict the flow behaviour using a computer, and the fluid never fails to agree with it. Repeat a laminar flow experiment and you will obtain a result identical to the first one. Flows at high speed are completely different. The fluid motion is three-dimensional, it keeps changing in time, and, vexingly, the results are unpredictable. If you were to carry out ten different experiments for a high-speed pipe flow, the ten results would be almost the same but not quite. Carry out a thousand experiments, and you would get a thousand almost-identical-but-not-quite-the-same results. The flows in that case can only be described in a statistical, average fashion, but the results from individual experiments are unpredictable, even by the fastest computer available today.

This is not just a laboratory construct. Turbulent flows happen to be everywhere around us. The flow in the heart pumping your blood right now is turbulent, and so are blood flows in your large arteries. In fact, if you use a straw to drink soda from a can, chances are the soda will reach your mouth in a turbulent state. Air flow is also often turbulent, as can be visualized using cigarette smoke. The flow around aeroplanes is turbulent, which has been historically one of the reasons why so much research effort has been devoted by industry and academics to understanding turbulent flows. So in our everyday life, turbulence is not the exception but is very much the norm.

The difference between a laminar flow and the turbulent regime is one of the most fundamental and well-studied aspects of fluid mechanics, with over 80 years of research—but with early sketches turbulent flows famously found as far back as the early 16th century in the notebooks of Leonardo da Vinci. Turbulence is also extremely challenging, and it is fair to say that even some basic features of turbulent flows still remain poorly understood. Computers have been helpful in shedding light on the physics of turbulence, but nobody has yet been able to predict, for example, the equivalent to the Hagen–Poiseuille law for turbulent flows. A lot of progress has been made on the transition from a laminar to a turbulent state and the first appearance of turbulent behaviour, but much remains to be done.

Industrial pipes

Beyond the natural world, flows in pipes with a wide range of length scales are routinely used in industry. On one end of the spectrum, very small pipes have been involved in the active area of microfluidics. Motivated in part by the need to carry out high-resolution chemical analysis on small amounts of biological liquids, the field of microfluidics has emerged in the early 1990s as a new technological marvel. Microfluidic devices typically consist of networks of small pipes a few micrometres in width and up to millimetres in length, made of glass or of soft elastomers bound on a glass surface. We illustrate in Figure 16(a) three examples of such devices.

With tremendous impact in a variety of scientific domains, ranging from protein synthesis to the design of fuel cells, the development of microfluidics and lab-on-a-chip devices has been accompanied by the need to understand the fluid mechanics of small-scale pipe networks, where the flow is always laminar. This has led to many new discoveries. For example, scientists can now use these networks to dilute, concentrate, or mix flowing chemical solutions in a precise way, and carry out multiple similar operations in parallel. They can also use the properties of pipe flows to sort particles in flowing suspensions, for example platelets or red blood cells within a blood sample.

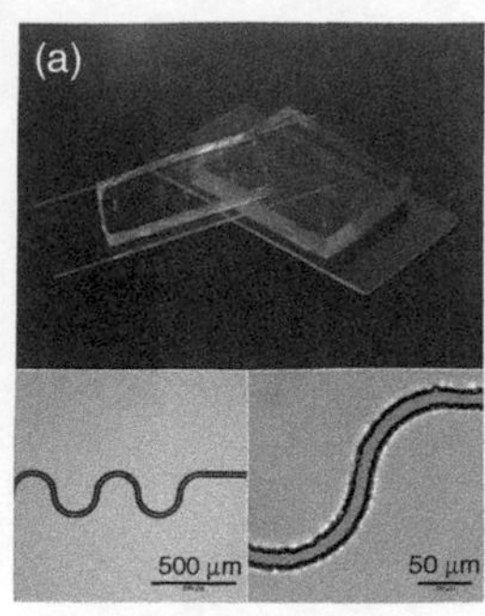

16. Flows in pipes are used in industry on a wide range of length scales. On small scales, serpentine channels in a microfluidic device are carved inside a transparent elastomer mould and glued to glass ((a) scale bars indicated in figure). At the other end of the spectrum is the Trans-Alaska oil pipeline near Prudhoe Bay, Alaska ((b) diameter of the pipe: 1.22 m).

At the opposite end of the spectrum, big pipes are used in the energy industry to transport gas, crude oil, and various petroleum products, sometimes over very long distances. The longest pipe in the world is the West–East gas pipeline in China, almost 9,000 km in length. The United States alone are covered by over 2 million kilometres of pipelines, an example of which is the Trans-Alaska oil pipeline, some 1,300 km in length and shown in Figure 16(b).

Liquid and gas flows in these large pipes are always turbulent, and so it is challenging to predict the pressures required to drive flow at the required speeds, and therefore to estimate the associated energy costs. Important work in this area was carried out by the 19th-century French hydraulic engineer Henry Darcy. Also known for his work on flows in porous media, Darcy installed a famous fresh-water piping system in the city of Dijon driven solely by gravity and requiring no pumps. As part of that work, he proposed an empirical equation extending the Hagen–Poiseuille law to all flow speeds. His model was later improved upon by a contemporary, the German mathematician Julius Weisbach, and is now known as the Darcy–Weisbach equation. The important role of roughness for industrial pipes was later included by the American engineer Lewis Moody in the 20th century; we now refer to the chart showing the value of the empirical parameters in the Darcy–Weisbach equation as a function of the flow speed and the pipe roughness as the Moody diagram.

Chapter 4
Dimensions

In Chapter 3, we saw that pressure-driven flow in a pipe could display different physical regimes for increasing fluid speeds, from smooth and laminar to unpredictable and turbulent. But one might wonder which parameters of the pipe flow actually control the transition. Is it the viscosity of the fluid? Is it the applied pressure? How about the size of the pipe? Clearly, different ways to increase the speed of the fluid exist, and therefore it appears difficult to predict the relative importance of the different control parameters of the problem. When many things matter, which are the ones that *truly* matter?

As it turns out, only one number controls the transition to turbulence. The catch is that this dimensionless number, which we see below is called the Reynolds number, is itself composed of all the other control parameters of the problem. To understand its origin, we have to take a step back and consider the role that physical dimensions and dimensional analysis play in physics. This will allow us to then discover a whole suite of special dimensionless numbers governing the dynamics of fluids.

Physical dimensions

All physical quantities have dimensions. So far in the book we have talked about velocities, accelerations, forces, and stresses, whose hysical dimensions can be expressed using the three fundamental dimensions of mass (M), length (L), and time (T). A length is a length. A velocity (or a speed), V, is a length divided by a time, which we write as $[V] = L/T$, where we use brackets to indicate the dimensions of the quantity in question. An acceleration, a, is the rate of change of a velocity; hence its dimensions are $[a] = L/(T^2)$. As dictated by Newton's second law of motion, a force, f, is a mass times an acceleration, so its dimensions are given by $[f] = ML/(T^2)$. Consequently, because a stress σ is a force per unit of surface area, we can equate $[\sigma] = [f]/(L^2)$; hence the dimensions of a stress are $[\sigma] = M/(LT^2)$.

Physical dimensions are related to units, but they are not the same concepts. For example, many different units may be used to measure lengths: metres, yards, feet, inches, etc. Changing the unit means changing the way in which we decide to represent the physical quantity in question, but it does not change its intrinsic value.

Why should we care about physical dimensions? The answer came from the French scientist Joseph Fourier. Famous for numerous contributions to mathematics (chiefly what we now call Fourier series and Fourier transforms) and physics (notably his derivation of the heat equation), he brought forward an important notion about physical dimensions in his 1822 book on the theory of heat. His idea, now known as the concept of *dimensional homogeneity*, states that in any scientific equation, terms on either side of the equal sign need to have the same physical dimensions. So an apple cannot be equal to an orange.

Dimensional analysis

The concept of dimensional homogeneity might appear to be an intuitive idea, but it has remarkable consequences. It means, for example, that we can only add, or subtract, quantities with the same physical dimensions. If an equation involves the sum of many terms, every term in the sum needs to have the same dimension. Furthermore, when we try to determine the complex relationship between some control parameters and a quantity of interest, the control parameters can only be combined in a way that is dimensionally consistent with the measured quantity. As an illustration, when we consider the Hagen–Poiseuille law, Equation (4), the many parameters that govern the fluid flow are arranged on the right-hand side of the equation in such a way that the resulting formula has dimensions of a flow rate (i.e. L^3/T).

This idea can be pushed further and can sometimes be used to (almost) solve problems without any knowledge of the underlying equations. Consider Newton's shear experiment from Chapter 2, in which the goal is to compute the force acting on the plate per unit of surface area, F/A, as a function of the fluid and the geometry of the experiment. With two plates separated by a distance h and moving with relative speed U, the force exerted by the fluid per unit of surface area of the plate depends on only three parameters: the viscosity of the fluid μ, U, and h (i.e. we assume that A is large enough for the ratio F/A to not depend on A). The force per unit of surface area has dimensions $[F/A] = M/(LT^2)$, and it turns out that there is only one way to combine the parameters μ, U, and h to obtain the same dimensions. Indeed, we note that only the the viscosity includes the dimension of mass, with $[\mu] = M/(LT)$, so the force per unit of surface area must necessarily be proportional to μ and with no additional dependence on the viscosity; if not, this would violate Fourier's principle. Furthermore, the viscosity also has the same $1/L$ dependence as F/A so we need to combine the remaining two parameters, U and h, in order to make an inverse time, and the only way to do this is to divide U by h. Therefore, using only dimensional homogeneity, we obtain that necessarily

$$\frac{F}{A} = \mu \frac{U}{h} K, \tag{7}$$

where K is a constant with no dimensions. That is the only combination of μ, U, and h that works.

The power of the dimensional arguments we used is evident. Without knowledge of any of the underlying equations or physical mechanisms at play, and using only Fourier's concept of dimensional homogeneity, we were able to find the value of the force per unit area up to the knowledge of a constant. Importantly, since K is equal to the ratio between two terms of equal dimensions, it itself has no dimensions and thus is said to be dimensionless. Of course these dimensional arguments cannot completely solve the problem, and in order to determine the value of K we would either need to carry out an experiment (and a single experiment is sufficient since the value of K is universal) or we would need to solve the equations of fluid mechanics (which would lead to the classical value $K = 1$, see Newton's law of viscosity in Equation (2)).

The impact of dimensional homogeneity on equations was pioneered by the 19th-century Scottish scientist James Clerk Maxwell, one of the most celebrated physicists of all time. Maxwell's name is most famously associated with the equations of electromagnetism and the physics of gases. His work in dimensional analysis included his proposal of the three fundamental 'mechanical' dimensions M, L, and T from which others are 'derived'. Other fundamental physical dimensions exist, of course, including temperature and electric charge.

Dimensionless numbers

In a vast array of problems in physics, instead of simplifying the problem down to a single constant, some irreducible parameters always remain, called dimensionless numbers. These turn out to play a crucial role in our understanding of fluid mechanics.

To explain this, let's revisit the example of the flow in a pipe discussed in Chapter 3. The pipe has length L and diameter D and we impose the flow rate Q, which is equivalent to imposing the mean flow speed U, since $U = 4Q/(\pi D^2)$. We would like to calculate the change in pressure Δp between the inlet and outlet of the pipe. If the pipe is long, the relevant quantity is the change in pressure per unit length of the pipe, since in the laminar regime the fluid pressure varies linearly along the pipe length. We therefore consider a long pipe for which we wish to compute the ratio $\Delta p/L$. The pressure drop per unit length could be a function of the diameter of the pipe D, the material properties of the fluid (namely its mass density ρ and viscosity μ), and of course the mean speed U of the fluid in the pipe. Consequently, we are trying to determine how the quantity $\Delta p/L$ depends on the parameters ρ, μ, D, and U.

The way to proceed consists in cancelling out dimensions one by one until all that is left is dimensionless. The quantity we are aiming to determine, the pressure drop per unit length, has dimensions $[\Delta p/L] = M/(L^2T^2)$, so if we divide it by a combination of the parameters with the same dimensions we obtain a dimensionless number. Using the Hagen–Poiseuille law, Equation (4), we see that a combination of the parameters with the same dimensions as $\Delta p/L$ is $\mu Q/(D^4)$ and therefore $\mu U/D^2$. The dimensionless version of the pressure drop per unit length is therefore written as the ratio $(\Delta p D^2)/(\mu U L)$, which we will denote $\overline{\Delta p}$ for simplicity.

The key insight to proceed is due to the famed British scientist Lord Rayleigh, born John William Strutt. Winner of the Nobel Prize in Physics in 1909, Rayleigh impacted countless branches of physics, including acoustics, optics, and fluid mechanics. His intuition about dimensional homogeneity led to a very important point. Consider the dimensionless pressure gradient, $\overline{\Delta p}$, which depends on four parameters, all of which have physical dimensions (ρ, μ, D, and U). Rayleigh's key observation is that a dimensionless quantity is allowed to only be a function of dimensionless combinations of other physical parameters. If not, then the value of the dimensionless quantity of interest (here, $\overline{\Delta p}$) will necessarily depend on the unit system in which we choose to write our parameters... which cannot happen if at the end we want to end up with a dimensionless number. Consequently, we need to get rid of all physical dimensions by combining parameters in a way that 'cancels out' their dimensions; if this is not possible, then these parameters cannot be part of the solution.

In the case of pipe flow, this can be done in the following way. Of the four parameters on which the dimensionless flow rate $\overline{\Delta p}$ could depend, i.e. ρ, μ, D, and U, only two have dimensions that include a mass. They are the mass density, with $[\rho] = M/(L^3)$, and the viscosity, with $[\mu] = M/(LT)$. In order for the dimension M to no longer appear, $\overline{\Delta p}$ can only depend on the ratio between these two parameters, and thus $\overline{\Delta p}$ can in fact at most depend on the three parameters ρ/μ, D, and U. We can next carry out a similar procedure for the dimension of time. The ratio between density and viscosity has dimensions $[\rho/\mu] = T/(L^2)$, while the velocity U is the only remaining parameter including dimensions of time, since $[U] = L/T$. Therefore, to no longer have a dependence on time, we need to multiply both terms, so that in fact $\overline{\Delta p}$ can at most depend on two parameters: $\rho U/\mu$ and D. Finally, since we have $[\rho U/\mu] = 1/L$ and $[D] = L$, in order to remove the dimensions of length these two parameters must necessarily be multiplied. Our complicated initial problem can therefore be finally stated in a remarkably simple dimensionless form as

$$\frac{\Delta p D^2}{\mu U L} \text{ depends only on } \frac{\rho U D}{\mu}. \tag{8}$$

The beauty of the final result in Equation (8) is that it is much simpler than the original formulation of the physical problem. The pressure difference between the two ends the pipe was thought to depend on four different possible physical parameters, which would make it a very complicated problem. However, it turns out that the non-dimensionalized pressure drop only depends on a single combination of all the physical parameters, namely the ratio $\rho UD/\mu$. So in this case, only one thing truly matters.

Building on the pioneering work of Fourier, Maxwell, and Rayleigh, this methodology of dimensional analysis can be used to systematically reduce any physical problem to one whose results depend solely on numbers that have no intrinsic dimensions, called dimensionless numbers. Since there exist three fundamental mechanical dimensions (M, L, and T), in general this allows us to reduce by three the number relevant parameters in the problem, as illustrated by going from the four-dimensional parameter in the original problem statement for the pipe flow to the single dimensionless number in Equation (8).

Reynolds number

The control parameter appearing on the right-hand side of Equation (8) is the Reynolds number, arguably the most important dimensionless number in all of fluid mechanics. For a fluid of density ρ and viscosity μ, a flow occurring with speed U on a relevant length scale D has a Reynolds number defined as

$$\mathrm{Re} = \frac{\rho U D}{\mu}. \tag{9}$$

In his original 1883 paper on the transition to turbulence, Reynolds suspected that the appearance of turbulent eddies in the pipe was governed by the combination of parameters in Equation (9), which explains why his name is now associated with it.

The Reynolds number is *the* control parameter used to classify a wide range of flow features, in particular the transition to turbulence in the pipe. Low-Reynolds number flows are smooth, laminar and fairly well-understood. In contrast, high-Reynolds number flows are turbulent, erratic, and much harder to predict. Since the pressure drop in a pipe flow depends on the value of the Reynolds number, so does the skin friction exerted on the pipe and so do the parameters in the Darcy–Weisbach equation from Chapter 3.

Because the Reynolds number is the one parameter ruling the dynamic of viscous fluids, it governs many aspects of our daily life. The Reynolds number sets the drag experienced by any vehicle moving in a fluid, such as a submarine in water or an aeroplane in air. It controls how a sailing ship can use the wind for its motion but also the forces experienced by trees in the same wind, and how fast leaves will fall on the ground. The Reynolds number determines the correct working regime of a combustion engine but also how fast it takes for cigarette smoke to disappear from a room. In fact, the Reynolds number is so omnipresent in fluid mechanics that it is hard to think of a real-life situation where it does not matter.

Womersley number

Many other dimensionless numbers can be used to characterize the variety of ways in which fluids can move. For example, instead of being steady, the flow in Reynolds' pipe could be made to oscillate in time with frequency f. A situation where time-periodic oscillations are significant is the flow in the circulation, where blood is periodically pumped by the beating heart in our network of veins and arteries. In that case, the dimensions of the frequency $[f] = 1/T$ provide a different method to cancel out the dimensions of time appearing in the Reynolds number. Since the ratio between density and viscosity has dimensions $[\rho/\mu] = T/(L^2)$, we could instead make the dimension of time disappear by considering the product $\rho f/\mu$, which has dimensions $[\rho f/\mu] = 1/(L^2)$. In a pipe of diameter D, the term $\rho D^2 f/\mu$ is therefore dimensionless. In the context of blood flow, this leads to the dimensionless Womersley number

$$\mathrm{Wo} = \sqrt{\frac{\rho D^2 f}{\mu}}, \tag{10}$$

named after the 20th-century British mathematician John Womersley.

The Womersley number captures the distribution of flow inside a blood vessel, as seen from patient data reproduced in Figure 17. Blood flowing in small vessels have low or moderate Womersley numbers and flow speeds varying parabolically across the vessel (Figure 17, brachial artery with Wo ≈ 3). In contrast, blood flows in large arteries at large Womersley numbers have speeds characterized by a very flat variation across the vessel (Figure 17, abdominal aorta with Wo ≈ 13).

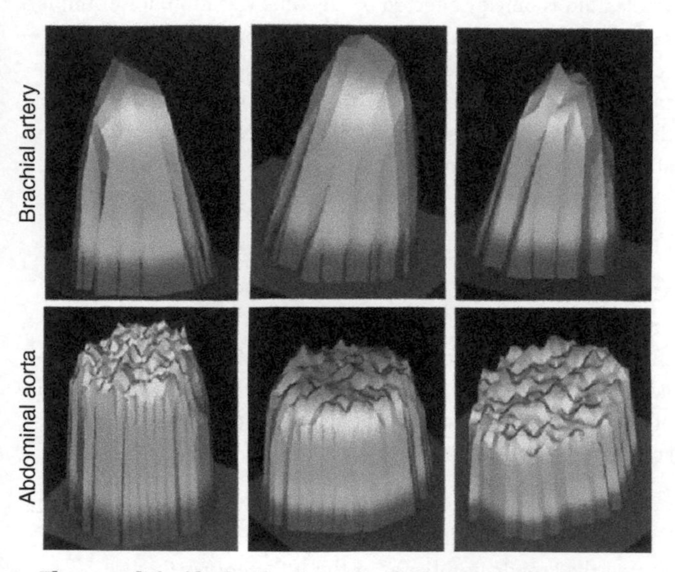

17. Flow speeds inside the blood vessels of six patients. The flow is moving upward on the figure, and the peaks indicate the magnitude of flow speed, with zero flow on the vessel walls and maximum flow near the centre of the blood vessel. The distribution of flow speeds in the vessels depends strongly on the Womersley number. In a brachial artery, flow occurs at moderate Wo ≈ 3 and the flow speeds vary parabolically through the vessel. In contrast, in the larger abdominal aorta, the Womersley number is large (Wo ≈ 13) and the flow speeds are much more uniform in the vessel.

A physical interpretation of the Womersley number in Equation (10) can be offered by rewriting it as a ratio of length scales,

$$\mathrm{Wo} = \frac{D}{\ell}, \quad \ell = \sqrt{\frac{\mu}{\rho f}}. \tag{11}$$

The length ℓ is the typical scale over which a periodic perturbation of frequency f is able to propagate into the viscous fluid in the direction perpendicular to the flow direction. Flows with low Womersley numbers have D small compared to ℓ and thus the whole fluid is quickly affected by any time variation, for example a variation of the blood pressure (Figure 17, brachial artery). In contrast, when the Womersley number is large, the value of ℓ is small compared to D, so a large portion of the fluid does not feel the imposed time variations and tends to respond with an almost-uniform flow magnitude (Figure 17, abdominal aorta).

Froude number

Many flows in our everyday life are affected by gravity, from drinking and crying to raining and flooding. The dimensions of the acceleration due to gravity are $[g] = L/(T^2)$, so in a geometry with a characteristic length scale D the product gD has dimensions of a square of a velocity. If the fluid has typical speed U, this suggests a dimensionless number

$$\mathrm{Fr} = \frac{U}{\sqrt{gD}}, \tag{12}$$

named the Froude number after the 19th-century English engineer William Froude. The Froude number measures the importance of gravity in affecting the dynamics of the flow. In flows with high Froude numbers, gravity does not play an important role while at low Froude numbers, gravity impacts the fluid dynamics, for example in the flows of rivers and oceans.

The intuition behind the Froude number is that it represents a ratio between two forms of energy. When a fluid particle is

moving, it carries with it some kinetic energy. If, during its moti
it is also subject to gravity, then it also possesses gravitational potential energy. These two forms of energy can be exchanged during the flow and the Froude number captures the typical r
between kinetic and gravitational energies. An illustrative ex
considers a large container of depth D emptying due to a gra
from a small hole at its base with speed U. A small fluid par
mass M has gravitational potential energy MgD at the top of the container (relative to its bottom), and almost zero kinetic energy if the container is large. When the fluid leaves the container it has kinetic energy $\frac{1}{2}MU^2$ and has lost all of its gravitational potential energy, so conservation of energy means $\frac{1}{2}MU^2 = MgD$, which can be rewritten using Equation (12) as $\mathrm{Fr} = \sqrt{2}$. The quiescent fluid in the container continuously trades all of its gravitational potential energy into kinetic energy of the flow, and the Froude number captures this balance.

A situation where the Froude number governs the fluid dynamics is the phenomenon of hydraulic jumps occurring in numerous natural and industrial settings. In this setup, famously first reported by Leonardo da Vinci, a shallow high-velocity flow transitions abruptly ('jumps') to a different regime with a thicker fluid layer and much lower speed. This is illustrated in Figure 18 for flow in a kitchen sink (a) and at the Burdekin Dam in Australia (b). As the fast flow, characterized by $\mathrm{Fr} > 1$, decelerates, the Froude number progressively decreases and when it reaches the critical value of $\mathrm{Fr} = 1$, the hydraulic jump occurs: a sudden rise in the surface of the liquid accompanied by an abrupt deceleration, after which the flow has $\mathrm{Fr} < 1$. During the jump, which is often associated with the creation of turbulence and mixing of the fluid, we have a transfer of kinetic energy from the flow into gravitational potential energy of the thicker but slower-moving flow.

Historically, the Froude number has played an important role in the design of ships as its value governs the wake and the waves

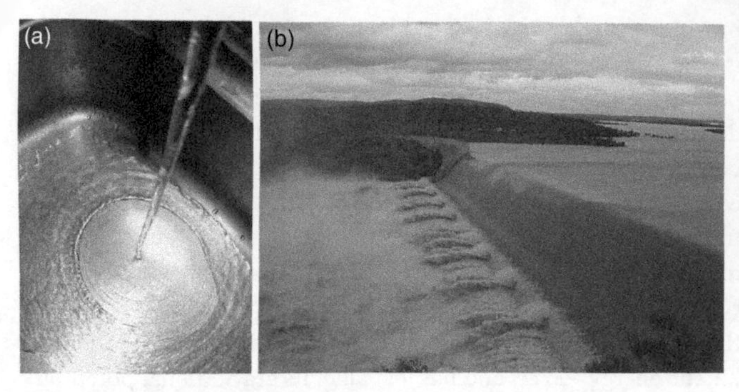

18. In a hydraulic jump, such as that seen in a kitchen sink (a) or at the Burdekin Dam in Australia (b), the fluid flow undergoes an abrupt deceleration and increases in thickness at a critical value of 1 for the Froude number, indicating an exchange between kinetic energy and gravitational potential energy.

observed on the water surface, which in turn impact the drag experienced by the ship. The classical work on wave drag is due to Lord Kelvin, born William Thomson, an Irish-Scottish physicist and one of the founding fathers of modern classical physics in the 19th century. Lord Kelvin demonstrated that, at small Froude numbers, the wake behind a body moving along a water surface makes a constant angle with the direction of motion, approximately equal to 19.5°. This is illustrated in Figure 19(a) in the case of a paddling duck. Recent work showed that the angle of the wake decreases when the Froude number exceeds a critical value of 0.5, as illustrated in Figure 19(b) for the wake behind a military ship.

Bond number

On small length scales, the Froude number tends to be small, but gravity can still play an important role—not relative to the kinetic energy in the moving fluid, but compared to the surface tension holding fluid interfaces together. Consider a liquid drop of size D and mass M in air. The typical gravitational potential energy in the fluid scales as MgD, and with the mass being the density ρ times

19. The surface wake behind a floating body depends on the value of the Froude number. At low Froude number, the angle between the wake and the body motion is constant and equal to Kelvin's prediction of approximately 19.5° ((a) wake behind a duck with Fr $\approx$ 0.15). When the Froude number exceeds a critical value of 0.5, the size of the wake decreases with the Froude number ((b) wake behind a military battleship with Fr $\approx$ 1.03).

the fluid volume, we have gravitational potential energy of order $\rho D^3 \times gD = \rho g D^4$. Since the surface tension γ is the energy associated with the air–fluid interface, the surface energy of the drop is of magnitude γD^2. The ratio between gravitational and surface energies is thus given by

$$\text{Bo} = \frac{\rho g D^4}{\gamma D^2} = \frac{\rho g D^2}{\gamma}, \tag{13}$$

a dimensionless group named the Bond number after the 20th-century English physicist Wilfrid Bond.

The impact of the Bond number on the shape of drops is illustrated in Figure 20(a) showing water droplets of various sizes deposited on a hydrophobic lotus leaf. Small droplets have a small Bond number so they are approximately spherical, which minimizes their surface energy. In contrast, big drops with large values of Bo are dominated by gravity, and thus they spread in the shape of thin, flat puddles, which minimizes their gravitational potential energy.

Further physical intuition can be gained on the Bond number by rewriting it as a ratio of length scales

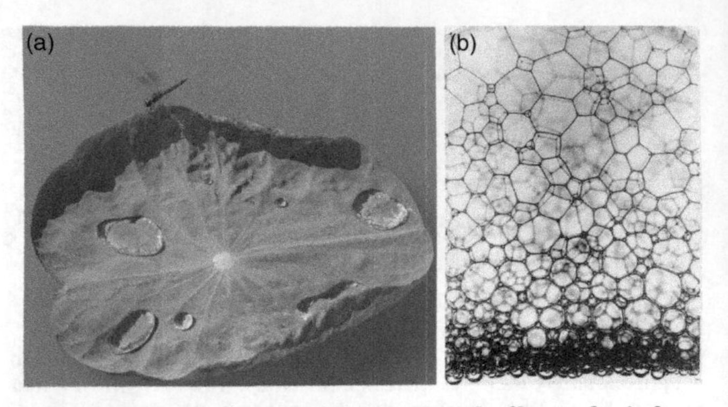

20. (a) Water droplets on a hydrophobic lotus leaf have spherical shapes at small Bond number but become flat puddles for large Bond numbers. (b) A liquid foam obtained by vigorously shaking a solution of soapy water. Most of the liquid eventually accumulates at the bottom where the bubbles are spherical, whereas those at the top are polyhedral.

$$\mathrm{Bo} = \left(\frac{D}{\ell_g}\right)^2, \quad \ell_g = \sqrt{\frac{\gamma}{\rho g}}, \tag{14}$$

with a new length scale ℓ_g, called the capillary length, which sets the critical size for drops to be in the low vs. high Bond number regime. For normal fluids on Earth, ℓ_g is on the order of millimetres, and therefore only sub-millimetric drops such as those found in the morning dew or on a windowpane remain approximately spherical. Of course, in situations where g becomes small, the capillary length can become much larger. For example, in the International Space Station the capillary length is larger than the station itself, and water drops of any size remain spherical if unperturbed.

The balance between surface tension and gravity has a famous application in the physics of foams. Well-known to all drinkers of beers and aficionados of bubble baths, foams are large collections of gas bubbles closely packed together in a fluid. Liquid foams are ubiquitous in cosmetic products such as shampoos and shaving creams, while solid foams are now used by the construction industry as insulating materials. An example of liquid foam obtained by shaking a solution of soapy water is illustrated in Figure 20(b), showing a collection of gas bubbles held together by surface tension and pushing against one another. Because of gravity, the liquid in a foam drains vertically downward. Bubbles at the bottom have therefore plenty of space to minimize their surface area and have spherical shapes. In contrast, bubbles at the very top have very little liquid left and tend to be polyhedral. The elongated liquid ridges where the bubbles intersect are called Plateau borders, and their junctions obey universal geometrical rules.

Weber and capillary numbers

In a fluid at rest, the Bond number captures the balance between gravity and surface tension. When the fluid is in motion, we need another dimensionless number to characterize the ability of flow to deform the fluid interface. For high-Reynolds number flows, this is captured by the Weber number, named after the German engineer Moritz Weber at the turn of the 20th century. Considering a fluid drop of size D moving with speed U, the kinetic energy of the drop is of magnitude $(1/2)MU^2$, i.e. $(1/2)\rho D^3 U^2$, while the surface energy is still γD^2. The Weber number We is then defined as (twice) the ratio between these two forms of energy

$$\mathrm{We} = \frac{\rho D^3 U^2}{\gamma D^2} = \frac{\rho D U^2}{\gamma}. \tag{15}$$

Given that a flow of air can exist around the fluid drop, we could also define a second Weber number using the density and speed of the gas flow (instead of that in the liquid).

Many natural and industrial flows occur at high Weber numbers, resulting in interfaces being stretched, deformed, and even broken. Perhaps the most famous example is that of raindrops, as shown in Figure 21(a). Drops fall down due to gravity; they accelerate and eventually reach a velocity that is high enough for the air flow to

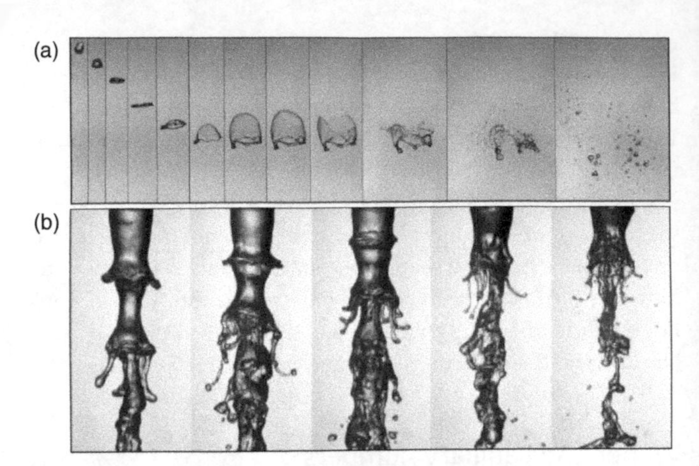

21. Fluid interfaces subject to high-Weber number flows get stretched and broken up into small droplets. This is the case for raindrops (a) and fluid jets subject to fast flow of air, leading to spray formation (b).

deform them significantly. What started as a compact water drop is stretched into a thin curved film, which eventually breaks into smaller drops. These drops will themselves also lead to new high-We flows when they fall into puddles and create splashes.

A related process occurring in industry is called atomization, whereby fluid jets are deformed by the fast flow of surrounding air (illustrated in Figure 21(b)). Here again, the Weber number is large so surface tension is not strong enough to hold together the liquid interface, which undergoes a cascade of instabilities until eventually breaking up and becoming a spray of small droplets. The atomization of liquid fuel is an important process in internal combustion engines, while a similar mechanism is also responsible for formation of sprays on the crests of breaking sea waves.

When drops are tiny and the flows around them occur at small Reynolds numbers, an alternative dimensionless number is used, called the capillary number Ca, and defined as the ratio $\mathrm{Ca} = \mathrm{We}/\mathrm{Re}$ so that

$$\mathrm{Ca} = \frac{\mu U}{\gamma}.$$

Instead of a ratio of energies, the capillary number n
interpreted as a ratio of forces, measuring the forces due
shear stresses deforming the drop relative to the resisting fo
from surface tension. Droplets at small capillary number are dominated by surface tension and therefore remain nearly spherical, while those with high values of Ca can undergo significant deformation, including break-up. Figure 22 shows the impact of the capillary number on the deformation of a single oil droplet in a controlled extensional flow ((a); see also Chapter 2). As the capillary number increases ((b); from top to bottom), the drop extends and becomes thinner. The capillary number plays an important role in many small-scale industrial situations where droplets are produced and manipulated, such as emulsions in the food industry (milk, ice cream) and many cosmetic products (creams, gels).

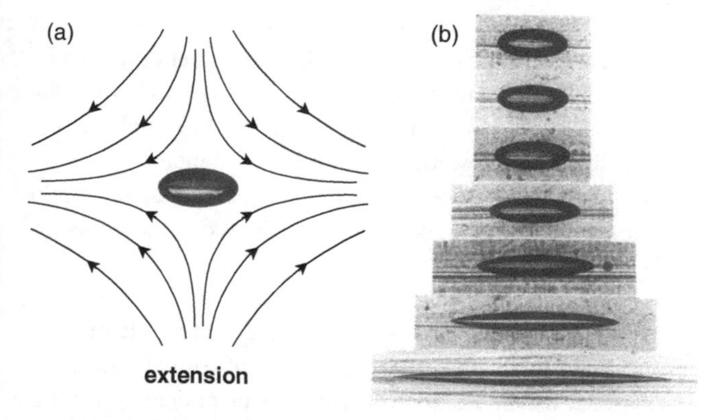

22. When a small drop is placed at the centre of an extensional flow, the viscous stresses stretch the drop and are resisted by surface tension (a). As the capillary number increases (b, from top to bottom) the magnitude of the drop deformation increases.

Mach number

While all other flows discussed in this book are incompressible, in some applications the effects of compressibility can become quite important. This is the case, for instance, for air flow around fast-moving bodies such as aircrafts and projectiles. The effect of compressibility is captured by the dimensionless Mach number, which compares the speed of the fluid flow U to speed of sound in air c,

$$\mathrm{Ma} = \frac{U}{c}, \tag{17}$$

and is named after the Austrian scientist Ernst Mach. In the atmosphere, the value of c depends on temperature and altitude, and ranges between 275 and 340 m/s.

Incompressible flows have very small values of Ma and the effects of compressibility start to be felt at $\mathrm{Ma} \approx 0.3$. When the Mach number increases, a sharp transition occurs near $\mathrm{Ma} = 1$, after which the surrounding flow is faster than the speed of sound and is said to be supersonic. This can happen in gas pipelines, where air can flow at speeds of hundreds of metres per second, and, of course, for flows around aeroplanes. Most of us have not experienced this transition ourselves since all current passenger planes have maximum flying speeds corresponding to $\mathrm{Ma} \approx 0.85$. In the 1960s, commercial aeroplanes breaking the sound barrier (i.e. exceeding $\mathrm{Ma} = 1$) were introduced, most famously the Concorde, now retired, which flew with a maximum Mach number of 2. The record remains $\mathrm{Ma} \approx 6.7$, also achieved in the 1960s by the US military.

When the flow becomes supersonic, an important new feature appears, called shocks. Physically, the speed of sound in the air is the speed at which perturbations in pressure or density to the air can propagate and go away from any disturbance. When the air flow becomes supersonic, the motion is itself faster than that propagation speed, and therefore none of the disturbances created

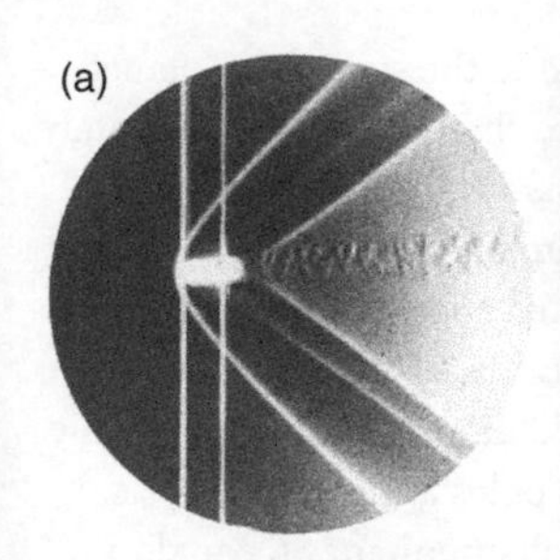

23. When a body moves through air faster than the speed of sound, a shock develops in a conical region behind the body where the fluid properties undergo very large changes over small distances. Shocks were first observed by Ernst Mach ((a) 1888) and can be seen around fighter jets as vapour clouds ((b) US Super Hornet fighter jet).

by the moving body is able to get out of the way. This leads to very large changes in pressure, velocity, and density over very small distances in the air flow (below one micrometre). Shocks were first observed by Mach; in Figure 23(a) we reproduce one of his famous 1888 photographs showing the shockwave around a supersonic bullet using a technique allowing to visualize changes in air density.

Shocks occurring in supersonic flows are also associated with loud noises called sonic booms, which propagate backward and can be heard in a conical region behind the moving body. An example of an everyday sonic boom is the crack of a whip, where the Mach number is about 2. Sonic booms can also be heard at air shows when military planes fly over the spectators. In that case, since shocks are also associated with sudden changes in temperature, water condensation leads to a thin cloud moving with the plane allowing us to observe the shock (the example of a US Super Hornet fighter jet is shown in Figure 23(b)).

Rossby number

A final dimensionless number allows us to capture large-scale flows affected by the rotation of the Earth. As both the oceans and the atmosphere are very thin compared to the size of our planet (a few hundred kilometres compared to the Earth's approximately 6,400 km radius), their flows occur mostly along the Earth's surface with very little motion in the direction perpendicular to it. But since our planet rotates about a fixed North–South axis, this means that the perceived value of the Earth's rotation, i.e. that projected along the direction perpendicular to the typical oceanic or atmospheric flow, is maximal at the poles and zero at the equator. This is quantified by the Coriolis frequency f, equal to twice (for mathematical convenience) the perceived rotation at any given latitude. In a domain of size D, the product Df has dimensions of a speed, so if a flow has typical speed U, this suggests a new dimensionless number as

$$\mathrm{Ro} = \frac{U}{Df}, \tag{18}$$

which is termed the Rossby number after the 20th-century Swedish-American meteorologist Carl-Gustav Rossby.

In situations where the Rossby number is large, the effects of the Earth's rotation are not important. This is the case at low latitudes (for example, at the equator, where f is zero, or in the tropics) or for very high flow speeds, for example in tornadoes. This can even be the relevant limit near the poles where the perceived Earth's rotation is the strongest, provided that the length scale D is small enough. For example, consider the famous case of a bathtub draining near one of the poles. Taking a typical bathtub size of $D \approx 1$ m and draining speed of $U \approx 1$ m/s, the rotation of the Earth being about 10^{-5} Hz leads to $\mathrm{Ro} \gg 1$. Without a remarkable control of the experimental conditions, it is therefore not possible for the Earth's rotation to affect the direction in which the fluid drains—that remains a hydrodynamic myth.

In contrast, when the Rossby number is of order one or less, the Earth's rotation plays an important role, as indeed is the case for many large-scale planetary atmospheric and oceanic flows. One of the more remarkable consequences of flows with small Rossby numbers are Taylor columns, named after the British physicist Geoffrey Ingram Taylor (more on him later). In a flow at low Rossby number, the fluid motion acquires a new rigidity in the direction of rotation so that a body moving in the rotating fluid appears to take with it an entire column of liquid above and below it as if it were a rigid cylinder spanning the whole fluid, and with no fluid motion in the direction along the axis of rotation. Perhaps the most famous (yet still controversial) Taylor column is the Great Spot in Jupiter's atmosphere.

Chapter 5
Boundary layers

As we saw in Chapter 2, the viscosity of a fluid produces fluid stresses in response to shear. In flows with high Reynolds numbers, shear rates can become large, with velocities changing substantially over small distances, thereby creating considerable stresses. A historically important discovery at the turn of the 20th century found that these large shear rates, occurring in regions called boundary layers, are not governed by geometry but instead occur spontaneously due to the physics of the moving fluid. To understand boundary layers, we have to go back in time to the world of fluid dynamics prior to their discovery and examine the subtle role played by viscosity when the Reynolds number is very large.

Drag and the limit of high Reynolds numbers

One of the main issues in the application of fluid mechanics to industry is the prediction of drag. Suppose a body moves through a fluid, for example an aeroplane through air or a submarine in the ocean. The drag is the magnitude of the force resisting the motion and acting in the direction opposite to the body's motion. Drag is important to industry because it is, ultimately, what governs economic cost; a significant portion of the airfare on your next plane ticket goes to the fuel needed to power the aeroplane against the air drag. A fundamental question is therefore: what is the value of the drag? It is clear that it depends on the Reynolds number, and in many situations relevant to industry, that number is large. In fact, it is exceedingly large; the Reynolds number of the air flow around a commercial aeroplane is in excess of tens of millions while around a moving submarine it can be hundreds of millions.

In the 19th century, computers were not available to help solve the complicated equations of fluid mechanics, and therefore scientists interested in estimating drag at high Reynolds numbers had to think. They came up with the following simple argument. Since the Reynolds number scales as the inverse of the viscosity, the large Reynolds number limit corresponds to a situation where viscosity plays a relatively small role in affecting the fluid motion. If that is the case, then surely the result for the flow must not be very different from that obtained by ignoring viscosity altogether. So scientists made the radical assumption to take the viscosity to be exactly zero; this is equivalent to saying that the Reynolds number is exactly infinite.

Perfect fluids

A fluid with zero viscosity is called a perfect fluid. If we get rid of viscosity completely, one term drops out of the Navier–Stokes equations governing the fluid motion (the second to last term in Equation (3)). In that case, they become the Euler equations, named after Leonhard Euler. The Euler equations are still very complicated but, under a wide range of assumptions, they can actually be solved exactly, and hence drag can be computed.

To solve the Euler equations, we have to assume the fluid flow has no swirling motion; using the terminology that will be introduced in Chapter 6, this is saying that the fluid has no vorticity. This is an assumption that turns out to be mostly adequate for flows away from any rigid surface. In that case, the solution of the equation is due to another 18th-century Swiss mathematician, Daniel Bernoulli. Born into a family of renowned mathematicians,

Bernoulli showed how to calculate the value of the pressure in the fluid explicitly. His challenging solution requires the use of a mathematical toolbox called vector calculus. We can, however, derive a simplified version of it, by only relying on the principle of conservation of energy in the fluid.

For example, consider the case of a cylindrical jet of fluid flowing at high Reynolds numbers and undergoing a change in its cross section. We ignore all effects due to gravity and assume that the flow is steady (i.e. not changing in time). The geometry is illustrated in Figure 24. Upstream in the jet of cross-sectional area S_1, the mean fluid speed is denoted by U_1 and the pressure by p_1; downstream where the area is now S_2, the speed and pressure are, respectively, U_2 and p_2. The flow is incompressible so the constant flow rate Q in the jet is given by $Q = S_1 U_1 = S_2 U_2$. Because the two speeds are different, the kinetic energy of the fluid changes between the upstream and downstream locations.

The principle of conservation of energy in the fluid states that the changes in kinetic energy of the fluid are due to the work done by the pressure forces on the fluid. The kinetic energy of a mass M of fluid is $(1/2)MU^2$. Since QT is the volume of fluid going through the surface during the time T, the amount of kinetic energy passing through any cross section during that time is given by $(1/2)\rho U^2 QT$.

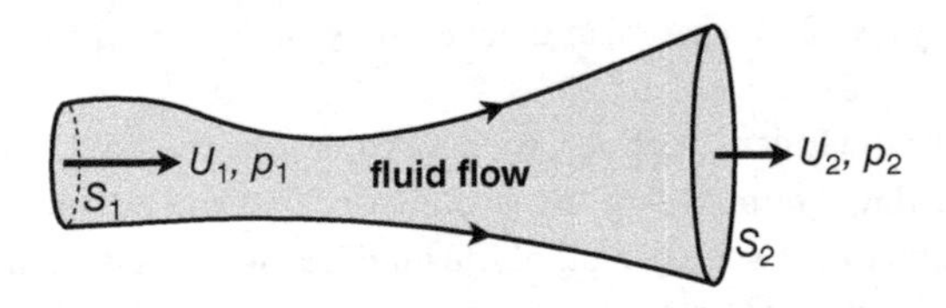

24. Applying the conservation of energy in an elongated fluid jet allows us to derive Bernoulli's equation. Upstream the jet has cross-sectional area S_1, the average speed is U_1 and the pressure is p_1. The corresponding values somewhere downstream are S_2, U_2, and p_2.

The force F exerted on the fluid due to the pressure is given by pressure times surface area, $F = pS$. When a force is exerted on a body moving a distance L, the work done by the force is equal to FL. In this case, during the time T the fluid moves a length $L = UT$ so the work done on the fluid by the pressure is equal to $FL = FUT = pSUT = pQT$. Conservation of energy is therefore written as

$$\frac{1}{2}\rho U_2^2 QT - \frac{1}{2}\rho U_1^2 QT = p_1 QT - p_2 QT \qquad (19)$$

or, after dividing by QT,

$$p_2 + \frac{1}{2}\rho U_2^2 = p_1 + \frac{1}{2}\rho U_1^2. \qquad (20)$$

The result in Equation (20), known as Bernoulli's equation, shows that the quantity $p + \rho U^2/2$ remains constant in the fluid; the same equation is obtained when solving the Euler equations directly. The physical intuition behind this result is that if $p_1 > p_2$ then the upstream pressure pushes on the fluid more, and therefore the fluid accelerates, resulting in $U_2 > U_1$.

D'Alembert's paradox

Bernoulli's equation shows that, provided that the quantity $p + \rho U^2/2$ is know somewhere in the fluid (and it is usually the case far from any disturbances), then wherever the speed of the fluid is known, so is the pressure, and therefore the drag on a rigid body can be evaluated. An early calculation of the drag was successfully carried out by the French mathematician Jean le Rond d'Alembert, famous for having edited the scientific portion of Diderot's multi-volume 18th-century *Encyclopaedia*.

The result from d'Alembert is both simple and profound. He showed that the drag on a rigid body moving at constant speed in a perfect fluid is always zero. So there is no drag whatsoever. This result is so surprising and so counter-intuitive to our daily life that it is now known as d'Alembert's paradox. Demonstrating this result rigorously is quite intricate mathematically, but we can gain the correct intuition by considering the flow of a perfect fluid past a spherical body.

Consider a sphere moving with steady speed through a perfect fluid. Moving with the sphere, we see that it experiences an incoming flow, and we illustrate its streamlines, i.e. the lines of flow, in Figure 25(a) (here again we assume that the fluid has no swirling motion). To calculate the drag, we need to know the traction on the surface of the sphere, i.e. the force per area exerted by the moving fluid. Since the fluid is perfect, its viscosity is zero, and therefore from Chapter 2 we know that the only traction is a normal stress due to pressure. Bernoulli's equation states that the quantity $p + \rho U^2/2$ is constant everywhere in the flow, so with the knowledge of U we can now obtain p and deduce the drag.

The key feature of the flow shown in Figure 25(b) is its perfect front–back symmetry. Specifically, the magnitude of the flow speed

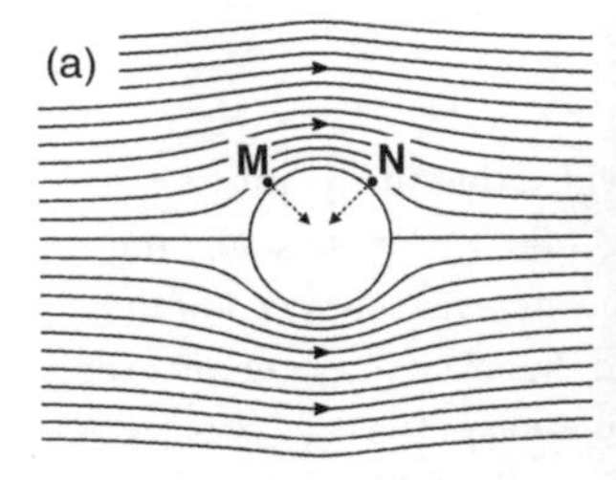

25. The flow past a rigid body at high Reynolds numbers was long thought to be similar to that obtained by neglecting viscosity. However, in the flow of fluid with zero viscosity ((a) illustrated on a cylinder), at two points M and N symmetric to one another, the fluid speeds are identical, so from Bernoulli's equation the pressures are the same, and therefore the total drag is zero. Prandtl showed that viscous stresses play a crucial role near rigid surfaces. The flow in (b), occurring at Re = 15,000 around a rigid sphere, is clearly qualitatively different from that in (a).

somewhere near the sphere in front of it is exactly the same as the symmetric point behind the sphere. Consider two points M and N, front–back symmetric to one another, as shown in the figure. The fluid speeds at M and N are the same, so following Bernoulli, we see that the pressures at these two points are also identical. At each point, the pressure exerts a stress that pushes against the sphere along the direction normal to the sphere (as indicated by arrows on Figure 25(a)). By symmetry we therefore see that the magnitude of the traction along the direction of motion at point M is equal and opposite to that at point N. But for every point on one side of the sphere we can find a symmetric point on the other side, so for every traction there is an equal and opposite traction on the other side as well. The total drag force on the sphere, which is the sum of all tractions, can therefore only be zero.

If the speed of the sphere is not constant but instead the body accelerates or decelerates, the fluid around it would have to also be accelerated or decelerated, which would result in a drag force (and the Bernoulli equation would have additional terms). But in the simple case where the body moves with a constant speed, the force agrees with d'Alembert's prediction of zero drag.

Clearly, this result is at odds with our everyday experience. Zero drag would mean no resistance from the surrounding air when we ride our bikes and no friction from the water when we swim in the pool. So what could have gone wrong? The problem turns out to hinge on the way the large Reynolds number limit was taken. Even for Reynolds numbers as large as can be imagined, the effects of viscosity somehow still matter. This is best illustrated in Figure 25 where we show experiments for the flow at high Reynolds numbers past a sphere and see that the fluid motion in the experiment (Figure 25(b)) is markedly different from the idealized perfect fluid result (Figure 25(a)). Why is it, then, that the mathematical case of zero viscosity is qualitatively different from that of a very large Reynolds number?

A revolution: boundary layers

In order to solve d'Alembert's paradox, a new way to find the approximate solutions of equations containing a very small parameter (the viscosity) had to be invented. The answer was a revolution, perhaps the biggest in the field of fluid mechanics, that came in a celebrated 1904 paper by the German engineer Ludwig Prandtl. Prandtl invented the concept of boundary layers; he explained how they appear in the fluid when the Reynolds number becomes very large and how much drag they produce.

As for many things in fluid mechanics, multiple ways to think about the intuition behind boundary layers exist. From a physical point of view, since the viscosity is responsible for shear stresses, assuming a vanishing viscosity means that one removes the source of friction in the fluid. However, that microscopic friction is responsible for enforcing the no-slip boundary condition on solid boundaries. Remove viscosity and you remove the origin of the no-slip condition, so the fluid will be allowed to slip on the surface of the body. A close inspection of the flow in Figure 25(a) does reveal slip occurring on the sphere surface. So from a physical standpoint, assuming a fluid to be perfect removes a crucial feature of fluid flow: its ability to enforce the correct (no-slip) boundary condition on rigid surfaces. It might be true that viscosity is not very important in most of the moving fluid, but near the surface of the body it matters tremendously.

Aside from the physics, an alternative mathematical explanation exists for the failure of the perfect fluid approximation. The Navier–Stokes equations satisfied by the fluid velocity and pressure are given in Equation (3). We do not analyse them in detail but one aspect is important. These are called differential equations, because they involve the spatial and temporal derivatives of our unknowns, specifically of second-order since they include two spatial derivatives of the fluid velocity. It is broadly known in mathematics that if one solves a differential equation with n derivatives, then one needs n boundary conditions for the problem to be well-posed. So to solve the second-order Navier–Stokes equations in the setup of Figure 25, we need two boundary conditions: one sets the speed of the fluid far from the body, and the other is the no-slip condition on the body itself. Now, if instead we assume the fluid to be perfect, the viscosity disappears and with it the term involving two spatial derivatives. We are then left with the first-order Euler equations involving a single spatial derivative and thus only enforcing a single boundary condition: we can match the correct flow at infinity but not the no-slip condition. This is the mathematical signature of the loss of friction discussed earlier.

In his landmark 1904 paper—which, more than 100 years later, is still the way we understand flows—Prandtl demonstrated that even in the limit of exceedingly high Reynolds numbers the viscosity always matters. However, the effect of viscosity is confined to thin 'boundary' layers near rigid surfaces. Outside these regions, the flow behaves, to a very good approximation, like a perfect fluid; inside the boundary layer things are very different and the flow has to be brought down quickly to zero on the surface to satisfy the no-slip condition.

The size of a boundary layer

One of the most important features of boundary layers is that they become thinner when the Reynolds number increases and can, in fact, quickly become much smaller than the body to which they are attached. For example, the boundary layer on an aeroplane wing at cruising speed can be smaller than a millimetre. To understand why this is so, we can consider the very simple case of a thin plate in a flow of speed U (Figure 26). We denote by L the length of the plate and by W its width and can use a physical argument to estimate the thickness of the boundary layer, denoted by δ.

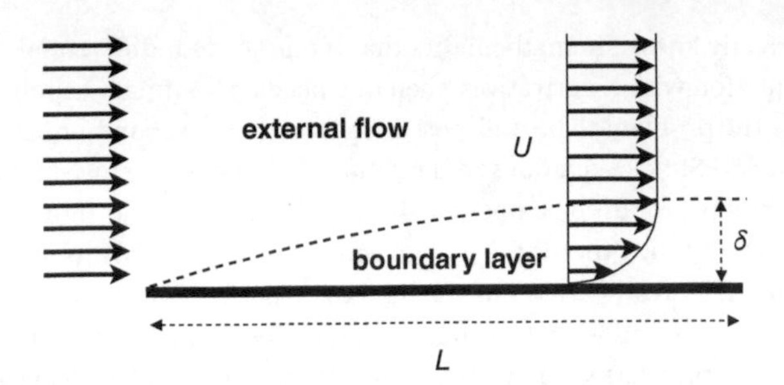

26. A fluid flows with speed U over a rigid plate of length L. When the Reynolds number is very high, the flow is brought to zero on the plate in a very thin region called a boundary layer. The thickness of the boundary layer, δ, decreases when the Reynolds number increases.

Newton's second law states that the rate at which momentum is lost by the incoming fluid is equal to the force resulting from the viscous stresses, all of which are concentrated in the boundary layer. The momentum in the boundary layer is on the order of MU where the mass M of the fluid in the boundary layer is $M \sim \rho V$ for a fluid volume of approximately $V \sim \delta LW$. So the momentum in the boundary layer has magnitude $\rho\delta LWU$. Driven by the flow at speed U, changes in the boundary layer along the plate occur on the flow time scale $T \sim L/U$ so the rate at which is the momentum is changing in the boundary layer is given by the ratio $(\rho\delta LWU)/T \sim \rho\delta WU^2$.

Inside the boundary layer, the viscous shear stress σ has a magnitude predicted by Newton's law of viscosity, Equation (2), as $\sigma \sim \mu U/\delta$, where δ is the thickness of the boundary layer and, therefore, approximately the length scale across which the flow near the plate changes from 0 to U (in other words, the shear rate in the boundary layer is approximately U/δ). Since a shear stress is a force per unit of surface area, the total viscous force exerted by the plate on the fluid has magnitude $F \sim \sigma LW \sim \mu LWU/\delta$.

With this, we can directly apply Newton's second law and balance the rate at which momentum is lost by the total viscous force so that

$$\rho W \delta U^2 \sim \mu LWU/\delta \tag{21}$$

which leads to an explicit estimate for the boundary layer size

$$\frac{\delta^2}{L^2} \sim \frac{\mu}{\rho UL}. \tag{22}$$

We recognize on the right-hand side of this estimate the inverse of the Reynolds number for the flow past the plate, and therefore we can write the size of the boundary layer as

$$\delta \sim \frac{L}{\sqrt{\mathrm{Re}}}. \tag{23}$$

This famous inverse square-root scaling shows that for large Reynolds numbers, the boundary layer is much thinner than the size of the plate, $\delta \ll L$. The fluid speeds near the plate therefore undergo changes over very small distances, which leads to considerable shear stresses and drag.

Flow separation and wake

In the boundary layer discussion above, we did not mention the flow pressure. The reason is that the pressure inside the boundary layer is always inherited from the pressure outside it. In the case of a flat plate, the flow outside the boundary layer has constant speed and constant pressure, so there is no pressure gradient there, and thus also no gradient inside the boundary layer. However, in flows around more complex shapes, the flows outside the boundary layer usually have gradients in pressure, which in turn affect the features of the flow inside the boundary layer. A prototypical shape that is not elongated in the flow direction is called a bluff body, for example a sphere—or a non-streamlined human being. In that case, the fluid dynamics inside the boundary layer becomes very subtle: the flow momentum changes not only because of viscous stresses but also because of the pressures inherited from the outside flow.

When the changes in pressure along the body are such that, on their own, they would induce fluid motion in the same direction as the actual flow (i.e. when the pressure decreases in the direction of the flow, similarly to the pipe flow of Chapter 3), we call them favourable; in contrast when they push the other way we referred to them as adverse. A favourable pressure gradient would be like blowing air with the wind while an adverse one like blowing against the wind. Complications arise every time an adverse pressure gradient exists in a boundary layer. Near the leading edge of the body (i.e. the tip of the body upstream of the flow), the boundary layer is thin, it remains along the body and the pressure gradient is favourable. If the pressure gradient stays favourable all along, then the boundary layer stays attached to the body. However, if at some point the pressure gradient becomes adverse, it will slow down the flow in the boundary layer. In fact, if it is sufficiently adverse for a long time, the flow comes to a halt and the boundary layer completely detaches from the body.

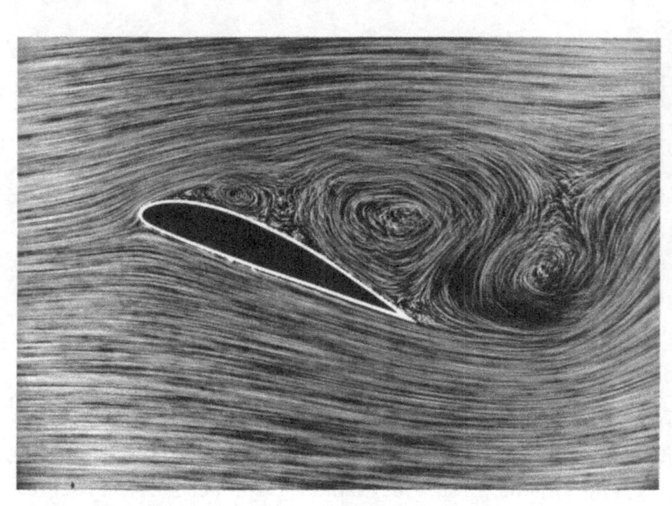

27. Air flow past an aeroplane wing allows us to illustrate the situation where the boundary layer remains attached to the body (region underneath the wing) and the case where the boundary layer separates from the body (region above the wing, leading to a wake of vortices).

This phenomenon, whereby a boundary layer stops and leaves the body, is called flow separation. We illustrate it in Figure 27 in the case of a flow past a tilted wing. The flow beneath the wing has not separated, and the boundary layer there remains attached throughout. In contrast, the flow above the wing separates very quickly, and after the separation point we see a wake of vortices (see Chapter 6). Separation can also be observed in Figure 25(b) for flow past a sphere at about the halfway point along the surface of the body.

Body shapes with little flow separation, or none at all, are called streamlined, with examples including many marine animals such as rays, eels, and most other fish. In industry, a century of aircraft design has produced wings whose shapes are streamlined (in an incoming flow aligned with the wing). We illustrated this in Figure 28 where numerical simulations of the flow past an Airbus A380 show the streamlines past the wing (i.e. the path of fluid

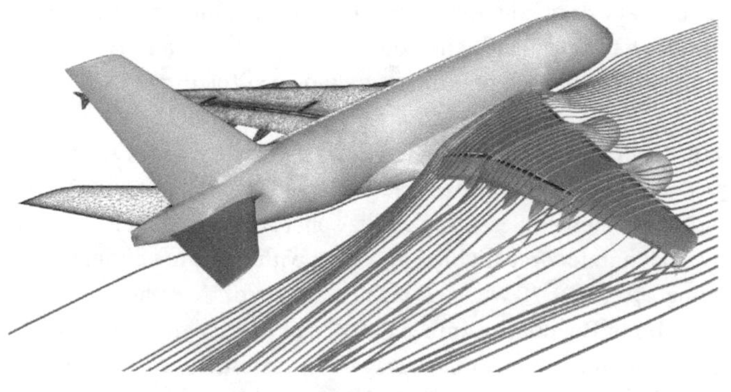

28. Computer simulations are used to model the air flow past an Airbus A380 using a mathematical grid illustrated on the left wing. The levels of grey on the plane fuselage shows the values of the fluid pressure. The lines over the right wing are flow streamlines indicating the instantaneous path of fluid particles in the unseparated boundary layer.

particles; the levels of grey on the body of the plane is the pressure field). Although the flow occurs at very high Reynolds numbers, the streamlines remain attached to the wing on their entire span. If separation were to occur, the appearance of the wake would be associated with a steep increase in the drag acting on the body. In some cases, this drag is beneficial. For example, aeroplane wings have flaps called spoilers that are lifted up immediately upon landing to induce separation and help slow down the plane.

Laminar vs. turbulent boundary layers

In Chapters 3 and 4, we saw how the Reynolds number controls the transition from laminar to turbulent flows. The same transition can also take place inside a boundary layer, resulting in even steeper shears and higher local drag. However, when the boundary layer becomes turbulent, the vigorous motion happening inside it delays flow separation. So an interesting compromise arises when it comes to drag at high Reynolds numbers: when the boundary layer flow becomes turbulent, the component of the drag due to the boundary layer goes up, but separation is delayed so the wake drag goes down. In some cases, inducing a transition to turbulent flow inside the boundary can in fact lead to a decrease of the overall drag, a phenomenon known as the drag crisis.

The drag crisis is illustrated in Figure 29 for the flow at speed U past a rigid sphere, whose surface is smooth (dark solid line). We plot the dimensionless drag on the sphere (i.e. the total hydrodynamic force divided by a term with the same dimensions, namely $\rho\pi(UD)^2/8$ where D is the diameter of the sphere) as a function of Reynolds number $\mathrm{Re} = \rho UD/\mu$. At high Reynolds numbers, the dimensionless drag is approximately constant until it suddenly decreases at a critical Reynolds number; the drag crisis results from the transition to turbulence in the boundary layer and the delay of the separation. If instead of being smooth the surface of the sphere is rough, the drag crisis occurs at a lower value of the Reynolds number, and it is less pronounced (illustrated by the

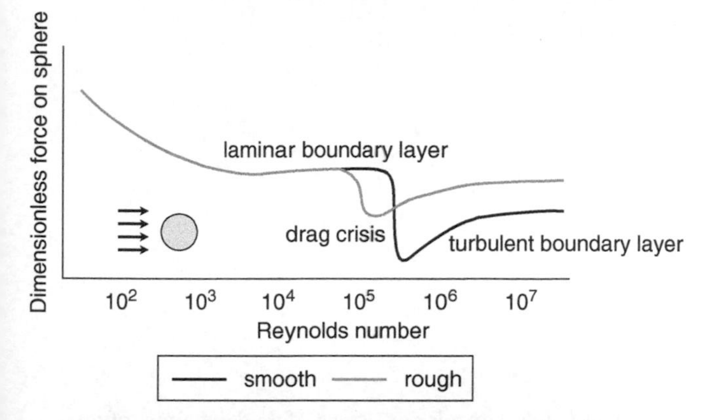

29. The dimensionless drag force experienced by a rigid sphere depends on the value of the Reynolds number. At high Reynolds numbers, the boundary layer is laminar but above a critical Reynolds number it becomes turbulent, delaying flow separation and leading to the drag crisis, a sudden drop in the drag force. This crisis happens at a smaller Reynolds number if the surface of the sphere is rough, and is less pronounced.

light solid line in the figure). The effect of roughness is to encourage the transition to turbulence to take place earlier, a useful feature in all applications where one wants to decrease drag. For example, small dimples on the surface of golf balls induce the drag crisis, thereby decreasing drag and allowing players to reach long distances. Other boundary layer control mechanisms developed in industry actively modulate drag, in particular on aircraft.

Chapter 6
Vortices

In this chapter, we consider an aspect of fluid dynamics often experienced in our lives, namely the localized swirling motion of a fluid, called a vortex. After considering some everyday examples, we will look at the concept of vorticity, which captures the tendency of fluids to rotate. The laws governing the dynamics of vorticity allow us then to understand the origin and transport of vortices and their applications in aviation and sports.

Everyday vortices

Our daily life presents many instances where the swirling motion of a fluid is concentrated in space, and thus vortices, also known as eddies, can be visualized. As we will see below, the spinning component of a flow can remain in the fluid for a long time, facilitating our experience of it.

The primary examples of vortices surrounding us are those in the weather system. We are all used to seeing weather maps on TV showing the rotating motion of large clouds, often accompanied by familiar terms such as cyclones and anticyclones. A cyclone is a large-scale rotational movement of the air in the atmosphere around a low pressure point, which is associated with inward air motion and wind. Its size can extend to thousands of kilometres, and the largest cyclones are typically found at the poles. Due to the rotation of the Earth, cyclones always rotate in the same direction: counter-clockwise (when seen from space) in the northern hemisphere and clockwise in the southern hemisphere. In contrast, in an anticyclone the rotation occurs around a point of high pressure, and in the opposite sense to cyclones in each hemisphere.

The cyclones that are often discussed in the news are tropical cyclones taking their strength from the evaporation of warm ocean water and whose strong winds can bring about devastation. These include hurricanes and typhoons, whose appellations vary depending on their location on the planet. An infamous example, illustrated in Figure 30(a), is the deadly 2003 Hurricane Isabel, whose winds reached 270 km/h over the Atlantic Ocean, leading to severe material damage and tens of deaths.

A related occurrence of a vortex, but a no less dangerous one, is a tornado, an example of which is shown in Figure 30(b). A tornado is a rotating mass of air that extends downward from the base of a big cloud. Tornadoes can be visualized from the ground because a cloud in the shape of a funnel is often located at the centre of it. The strong winds created by tornadoes are often very dangerous to humans; they can lift cars and destroy houses.

A smaller and safer version of a tornado can be generated in one's bathtub, or any time a large body of water is emptied from a small hole, such as in a kitchen sink. The origin of this so-called bathtub vortex lies in the quick amplification by the draining flow of any small initial rotation in the fluid, leading to the creation of a visible column of fluid and its interface with the surrounding air (see Figure 30(c)).

Another well-known vortical structure is called a vortex ring, akin to a tiny tornado looping back on itself. For example, vortex rings are produced by smokers who use their tongue and mouth to blow smoke in the shape of a doughnut. A similar form of smoke ring,

30. Everyday examples of vortices. On large scales, Hurricane Isabel can be seen from the International Space Station on 15 September 2003 (a) and a funnel cloud allows us to visualize a tornado in Manitoba, Canada (b). On smaller scales, the deformation of the interface between air and water is the signature of a bathtub vortex (c). Some vortices also take the shape of swirling rings looping back on themselves, for example the steam rings emitted in the air above Mount Etna (d) or water rings created by a beluga whale who, after blowing air in the vortex, plays with it (e).

also dangerous to human health but for different reasons, is the steam ring emitted by active volcanoes, such as the one above Mount Etna, shown in Figure 30(d). In water, similar ring-like vortices are created by dolphins and whales as toys (Figure 30(e)). They first create the ring vortex in the water using a quick motion of their fins, then they inject air into its centre by blowing in it. This allows them to visualize the water ring and encourages them to play with it.

Vortex shedding

A canonical case of swirling flow is the one where localized vortices are continuously injected into the fluid. For example, when a rigid body is placed in a uniform flow, the fluid flow remains laminar and steady up to a critical value of the Reynolds number, above which vortices are periodically shed from it. This flow is known as the von Kármán vortex street, named after the 20th-century Hungarian-American mathematician and engineer Theodore von Kármán. This phenomenon has been experienced by many of us who have had pieces of clothing flapping periodically in the wind, for example while riding a bike.

The simplest laboratory example of a vortex street occurs for a cylinder whose axis is perpendicular to the flow; in that case the critical Reynolds number is $\mathrm{Re} \approx 47$ using the diameter of the cylinder as length scale D in Equation (9). The appearance of the vortices behind a cylinder is illustrated in Figure 31(a). This periodic wake structure persists at very high Reynolds numbers, as shown in Figure 31(b) in the cloud dynamics of the atmosphere above the Robinson Crusoe Islands off the Chilean coast.

In a vortex street, since vortices are released periodically from the rigid body, a time scale appears in the problem, namely the frequency f at which the vortices are shed (f measures the number of events per second). Using the characteristic size D of the body and the magnitude of the incoming flow U, a dimensionless

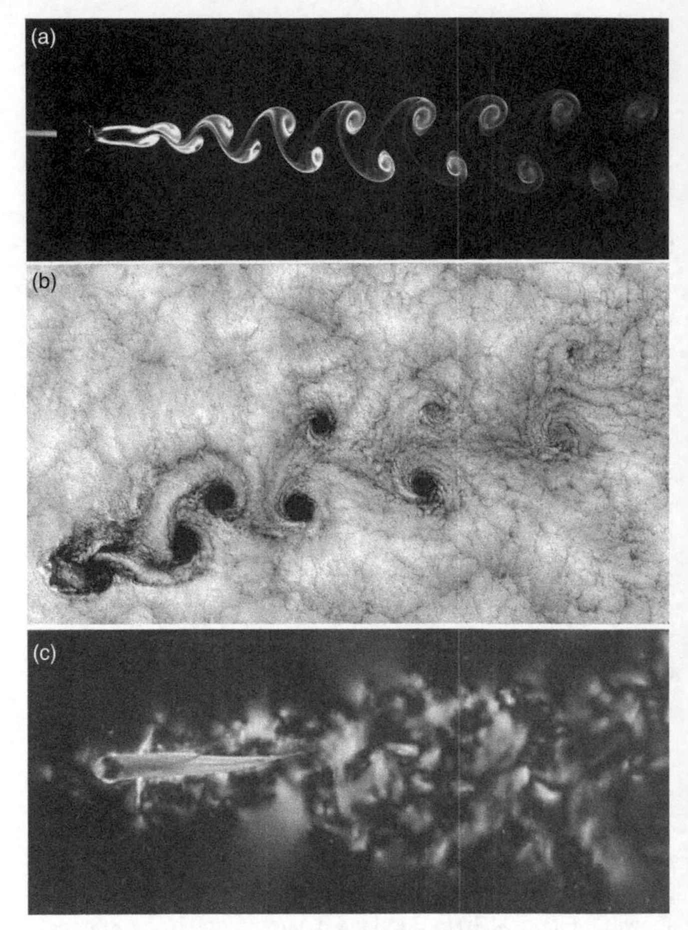

31. Vortices in wake flows. The ubiquitous von Kármán vortex street appearing in the flow past a rigid body can be observed in laboratory settings ((a): air flow behind a cylinder) or in the environment ((b): atmospheric air flow above the Robinson Crusoe Islands off the Chilean coast). Fish swimming in water generate a reverse vortex street in their wake, with eddies rotating in the direction opposite to the standard von Kármán wake (c).

number, called the Strouhal number, can therefore be defined as the ratio

$$\mathrm{St} = \frac{Df}{U}. \tag{24}$$

This dimensionless number is named after the 19th-century Czech experimental physicist Vincenc Strouhal, who studied the motion of thin wires oscillating in the wind.

Remarkably, over many orders of magnitude in the Reynolds numbers, the Strouhal number for a von Kármán vortex street remains approximately constant at $\mathrm{St} \approx 0.2$. Consequently, when the flow speed increases, the frequency at which the vortices are created increases approximately linearly with it.

The ubiquity of the von Kármán vortex street extends to the biological world, but with a twist. Many animals are known to self-propel in fluids, from fish and whales swimming in water to insects and birds flying in air. Flow measurements around these organisms have revealed that they also create von Kármán-like wakes, but with the intriguing property that the orientation of the vortices in the wake are reversed (see Figure 31(c)). All vortices that used to rotate in one direction for a standard von Kármán street rotate the other way near a self-propelling animal, and vice versa, so the wakes behind swimmers and flyers are often called reverse von Kármán streets.

The change in the direction of rotation of the vortices in the wake is associated with the fundamental role of fluid dynamics for propelling animals. In the case of a traditional von Kármán wake, the vortices appear in the flow as a result of the obstruction of the fluid by the body (e.g. the cylinder in Figure 31(a)). The vortices are therefore associated with drag on the body, and a loss of energy. In contrast, in a reverse von Kármán street, the vortices appear because the animal manipulates the fluid in order to create propulsion, and thus extracts some useful work from it. So the change in vortex rotation is intrinsically linked to the difference between drag (not helping) and propulsion (helping).

To further characterize the fluid mechanics behind an animal propelling itself in a fluid, a second Strouhal number can be defined. If we denote by A the amplitude of the periodic motion of the animal's appendage, for example the flapping of a bird wing or that of a fish fin, and if U is now the speed at which the animal moves, the new Strouhal number, $\widetilde{\mathrm{St}}$, is

$$\widetilde{\mathrm{St}} = \frac{Af}{U}. \tag{25}$$

Inspired by the discovery of an approximately constant Strouhal number for traditional von Kármán wakes, researchers have investigated what values of $\widetilde{\mathrm{St}}$ are applicable in the case of animals. As it turns out, essentially all known moving animals, swimming in water or flying in air, large or small, always move with a Strouhal number within a very narrow range of values, between about 0.2 and 0.4. This striking result suggests that animals have a universal way to create and manipulate vortices in order to extract useful propulsive work from them.

Vorticity

These examples of vortices all have in common that the flow undergoes a rotating motion concentrated in space. How do we think about and quantify this swirl? In the 18th-century pre-Prandtl world of fluid mechanics, Euler did realize that motions with local rotation of fluid elements could be included in the established theory of fluid motion. The word 'vorticity' was proposed to describe this local swirling motion by the British applied mathematician Horace Lamb, famous for his work on waves and as the author of influential textbooks at the turn of the 20th century.

To understand intuitively the concept of vorticity, the easiest is to picture a fluid flow with locally curved streamlines, as in

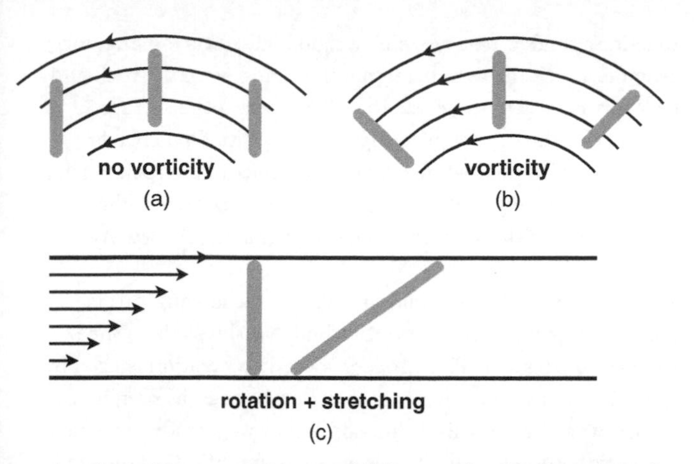

32. In a flow with no vorticity, material lines made of fluid particles do not rotate even when the streamlines are curved (a). In contrast, a flow with vorticity leads to the local rotation of the material lines (b). In a shear flow, material lines not only get rotated by the flow but they also get stretched (c).

Figure 32(a)–(b). The fact that the streamlines are curved does not necessarily mean that the fluid undergoes a local swirl. Imagine drawing a small straight line inside the fluid (shown in grey in Figure 32). This line is called a material line since it is composed of fluid particles, i.e. of the material. Now we can ask what happens to this material line when the flow moves it. Some flows transport the material line but do not change its orientation (Figure 32(a)). These are flows that do not have vorticity. In contrast, in a flow with vorticity, the material line rotates at it moves along the flow (Figure 32(b)). The vorticity, denoted by ω, is defined as twice the rate at which the material lines rotate locally. No rotation would mean zero vorticity and the faster this local rotational motion, the larger the vorticity.

However, why would a fluid have vorticity to begin with? In essence, the flow set up by any object moving in a fluid invariably includes some vorticity because of the no-slip boundary condition.

Imagine a fluid at rest in which a rigid body starts moving, for example a rock thrown in a pond. If we move with the rock and look very closely to it, we see that the flow near the surface resembles a shear flow, as in Figure 32(c), with fluid speeds increasing with the distance from the surface of the rock. Such a shear flow tilts any material line that is oriented perpendicular to the direction of the surface, which is a signature of vorticity.

Through the no-slip boundary condition, the moving surface therefore injects vorticity into the fluid. No-slip leads to shear, which leads to swirl. This process of vorticity creation is used by volcanoes and whales to create their vortex rings, as seen in Figure 30(d)–(e). Similarly, no-slip is how we introduce vorticity in the bath upon leaving it, leading to its amplification during draining and to the bathtub vortex. The no-slip boundary condition is also at the origin of the von Kármán vortex street from Figure 31 by forcing the fluid to come to rest on the stationary surface.

Destruction of vorticity

Once vorticity is in the fluid, how does it go away? Since viscosity is responsible for the creation of vorticity, it is only fair that it also acts as the main mechanism to make it disappear. Indeed, the same physical principle—the friction between fluid particles—governs both the creation and the destruction of vorticity. This friction allows the vorticity to be diluted in the fluid, and eventually to disappear, through a generic process known as diffusion.

To understand diffusion, one can use the analogy with the evolution of a dye in a fluid, e.g. a drop of food colouring introduced into water. Through diffusion, the initially concentrated colour moves slowly away from the initial point at which it was released, and by doing so it gets diluted and decreases in intensity. Inject a small amount of red dye in a swimming pool and eventually we will see no more red colour. There diffusion has a molecular origin and results from the random jittery motion of fluid molecules bumping into one another. The rate of diffusion is then governed by a key parameter known as the molecular diffusivity.

Diffusion is also the primary process to make vorticity vanish. Like a dye, any vorticity in the fluid is slowly diluted and eventually disappears. However, instead of molecular motion, the diffusion of vorticity originates from the viscous friction in the fluid flow. Thanks to viscosity, regions of fluid undergoing a local swirl are able to induce the rotation of nearby fluid particles, which themselves entrain their neighbours, etc. The rate of diffusion of vorticity is thus governed by the viscosity of the fluid. Specifically, vortices disappear quickly in fluids with high viscosity, for example honey, but are long-lived in low-viscosity fluids such as water.

Transport of vorticity

In between its creation and destruction, how does vorticity evolve in a fluid? How can we predict the dynamics of the vortices in Figures 30 and 31? The fundamental understanding of vorticity transport was laid out in a landmark 1858 paper by Hermann von Helmholtz. A German scientist and physician famous for his contributions to classical physics and neuroscience, Helmholtz explained in that paper how vorticity is moved and modified by a flow.

The transport of vorticity turns out to depend critically on the spatial dimensions of the flow. Flows can be either two- or three-dimensional. A flow is two-dimensional if no flow occurs in the third dimension and if no characteristics of the flow depends on that third direction; for example, a wind blowing near the earth with a direction and magnitude that is independent of the altitude. Many flows in the atmosphere and the ocean are approximately two-dimensional since their third dimension (the thickness of the atmosphere or the ocean depth) is much smaller than the other planetary scales.

For such two-dimensional flows, vorticity is moved precisely like a material line made of fluid particles. Just like a dye, vorticity is injected into the fluid, it is moved passively by the flow and eventually it diffuses away. Since no additional mechanisms modify them, vortices in these two-dimensional flows can be long-lived, for example explaining the long periods of time during which cyclones inflict damage.

If, instead, the flow is three-dimensional, Helmholtz showed that vorticity, again like a material line, can also be tilted and elongated by the fluid flow. Returning to the vertical material line element in the shear flow from Figure 32(c) we see that not only is it tilted by the flow, but it also increases in length. In the context of vortex motion, this elongation is called vortex stretching, an important mechanism to amplify vorticity. To understand vortex stretching intuitively we have to delve further into Helmholtz's contributions.

Vortex laws

In his 1858 paper Helmholtz derived three fundamental 'vortex laws'. He was interested in thin filaments where the swirl is very localized, for example the beluga whale vortex ring in Figure 30(e). Seen from sufficiently far away, the vortex ring appears as a slender structure of concentrated swirl, called a vortex filament, where the flow rotates locally around the direction of the thin filament. Helmholtz introduced the strength Γ of a vortex filament, also termed circulation by Kelvin. At a given location along the vortex filament, Γ is defined as the total integral value of the vorticity. Examining a vortex filament from up close we can clearly distinguish its non-zero thickness. In that case, it is sometimes called a vortex tube, with an illustration in Figure 33(a) showing the lines of vorticity and the local swirling motion of the fluid. Calling S the total cross-sectional area of the vortex and ω

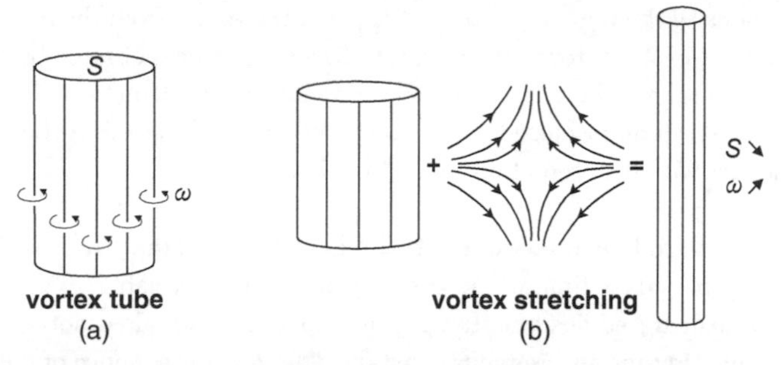

33. In a vortex tube, the flow in an area S rotates with vorticity ω around the direction of the tube (a). When a vortex tube is placed in an extension flow, it deforms like a material volume, so it gets contracted along one direction and extended in another (b). The cross-sectional area S decreases and thus, according to Helmholtz's third law, the vorticity ω has to increase. Vortex stretching leads therefore to an amplification of the vorticity.

the mean value of the vorticity there (i.e. twice the average rotation rate of material lines in the fluid), the strength of the filament is the product $\Gamma = S\omega$.

Helmholtz's first fundamental law states simply that the strength Γ of the vortex filament (i.e. its circulation) takes the same value at every point along the filament. This is a non-intuitive consequence of the fact that the vorticity is a mathematical object known as curl, a vector field with some special properties. As a result, the strength (or circulation) of the vortex filament is a single number characterizing it in its entirety.

Helmholtz's second vortex law, a consequence of the first, states that a vortex filament cannot end in a fluid. If it were ever to end in a fluid, the first law would not be satisfied since the strength would go abruptly from Γ to zero. The only way for a vortex filament to end is therefore either to extend to the boundaries of the fluid or to form a closed path. These two options are seen in the elongated vortices of Figure 30. Tornadoes extend from the clouds all the way to ground and the bathtub vortex from the free surface of the water to the plughole. The ring vortices emitted by the volcano and those created by the whale loop back on themselves along closed paths. So the structure of real-life vortices agrees with the predictions of Helmholtz.

In his third fundamental result, Helmholtz showed that, over short periods of time where viscosity does not appreciably dissipate the eddies, the strength of a vortex filament does not change in time and remains constant. This is a consequence of the conservation of angular momentum for the fluid since the strength of a vortex filament (or its circulation) is in fact proportional to the angular momentum per unit of mass. Beyond the case of vortex filaments, this result holds very generally, and is known as Kelvin's circulation theorem.

Vortex stretching

We can now predict the consequence of vortex stretching. Essentially, vortex stretching leads to an amplification of vorticity due to Helmholtz's third law. Consider a vortex tube as shown in Figure 33(a) where the flow has vorticity ω across an area S. If we place the vortex tube in an extension flow as in Figure 33(b), it behaves like a material volume and follows the motion of the fluid. Clearly the flow is going to compress the tube along the horizontal direction and extend it in the vertical direction and, as a result, the vortex tube becomes thinner and longer.

Helmholtz's third law states that, before viscosity has the time to act, the strength $\Gamma = S\omega$ of the vortex remains constant. The stretching of the vortex tube leads to a decrease of the cross-sectional area of the tube, S, and therefore the value of ω has to increase so that their product remains constant. An increase of the vorticity means that the fluid rotates locally with faster speeds. This amplification is therefore a direct consequence of the stretching of vortex lines by the flow.

Physically, this amplification of the swirling is due to the conservation of angular momentum. A famous illustration of this principle in a non-fluid mechanics context is used by spinning figure skaters who bring their arms closer to their body so as to increase their rate of rotation. The change in the position of the arms of a figure skater plays the same role as the decrease in the area of the vortex tube and both are accompanied by a faster rotation in order to conserve angular momentum.

The amplification of vorticity by extension flow plays an important role in the dynamics of tornadoes. In the summer, the air near the ground is warmed up by sunlight, so it becomes lighter and rises. Because of conservation of mass, if that air goes up vertically then some other air has to move horizontally in order to replenish it. The air flow near the ground is therefore similar to the extension flow from Figure 33(b), amplifying the rotation of any swirling flow and helping strengthen tornadoes. The same principle applies in the bathtub vortex of Figure 30(c). In that case, the drainage flow near the plughole is also a local extension and thus any remnant vorticity in the fluid gets strongly amplified, leading to a strong eddy.

Aeroplane vortices and lift

One everyday example where vortices are critical is the flight of aeroplanes. When a plane accelerates prior to taking off, the very sudden acceleration of its wings relative to the surrounding air leads to the creation, through the action of viscosity, of a large rotational structure called a starting vortex (see the illustration in Figure 34(a)). This eddy, which rotates in the counter-clockwise direction when observing the plane moving from right to left, remains on the runway until it eventually decays due to viscosity.

The starting vortex turns out to be the important physical ingredient allowing the plane to take off and fly, due to the laws

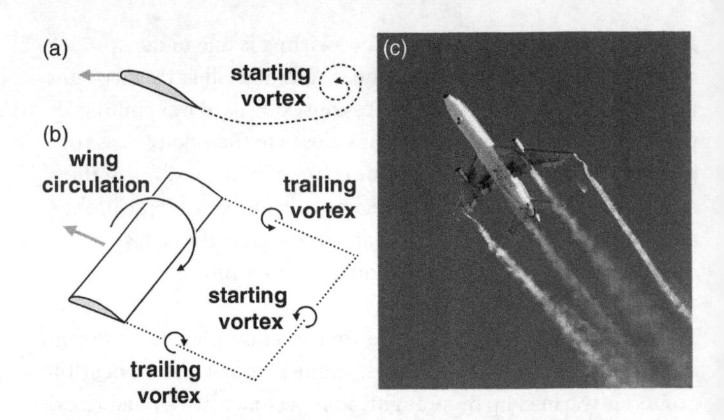

34. Vortices around aeroplane wings. When a plane takes off, each wing leaves on the ground a starting vortex, rotating in the counter-clockwise direction when observing the plane moving from right to left ((a) showing a cross section of the wing). Due to Helmholtz's third law, an equal and opposite flow rotation (i.e. in the clockwise direction) called the wing circulation is carried by the wing and is responsible for the lift on the plane (b). Vortex filaments, called trailing vortices and aligned with the direction of motion of the plane (b), are also created. These trailing vortices can be visualized using smoke generators ((c) here on a Lockheed L-1011 Tristar) or can often be seen naturally thanks to the condensation of water vapour.

derived by Helmholtz. Consider the air flow surrounding the entire aeroplane. Initially there is no motion, so the fluid is at rest and the vorticity is zero everywhere. Helmholtz's third law (or Kelvin's circulation theorem) says that the total circulation of the flow must remain equal to zero. However, because a starting vortex is left behind, it means that the wing must necessarily carry with it an equal and opposite vortex. Due to the starting vortex, a movement of air is always induced around the wing in the clockwise direction (when observing the wing going from right to left), as shown in Figure 34(b). This flow is called the wing circulation.

Wing circulation is what allows aeroplanes to fly through the generation of lift. The fluid mechanical forces exerted on bodies moving in fluid are of two types: drag and lift. Drag is the force in the direction opposite to motion, as discussed in Chapter 5. Lift, in contrast, is the force exerted by the moving fluid on the body in the direction perpendicular to the motion. Since an aeroplane aims to move horizontally, it needs lift in the vertical direction in order to balance its weight, allowing it to take off and fly. So the lift force exerted on aeroplane wings is the key feature allowing flight in the first place.

Why does wing circulation generate lift? We can understand it intuitively by examining the air flow around the wing during flight. Moving with the wing, we see that it experiences a strong air wind and can compare the speed of the wind above the wing (u_{top}) to the one below it (u_{below}). If the wing carries no circulation, we have approximately $u_{\text{top}} = u_{\text{below}}$, so by invoking Bernoulli's equation (see Equation (20)) the flow exerts identical pressure above and below, and therefore no net lift force is exerted on the wing.

Things are, however, very different for a wing with circulation. As illustrated in Figure 34, the effect of wing circulation is to induce an additional flow of air. Above the wing, that flow goes in the same direction as the outside wind while below the wing it goes in the opposite direction (i.e. against the wind). Comparing the total speed of the air flow above and below the wing we see that now $u_{\text{top}} > u_{\text{below}}$. If the air flow above the wing is faster than below, then Bernoulli's equation says that the flow pressure above the wing is smaller than that below it, i.e. $p_{\text{top}} < p_{\text{below}}$. This difference in pressure leads to a net force pushing vertically upward on the wing, in other words lift.

This mechanism of lift generation is subtle. First, a starting vortex has to be created to ensure that the wing carries with it an equal and opposite circulation. Second, that starting vortex needs to rotate in a specific direction. Indeed, a starting vortex rotating the other way than in Figure 34 changes the direction of the circulation around the wing, and the upward lift becomes a downward one with no hope for take off or flight. Third, the strength of starting vortex needs to be sufficiently large to generate a lift strong enough to balance the full weight of the plane. All these important properties turn out to be set by the shape and the size of the wing.

In addition to the starting vortex and wing circulation, the motion of aeroplane wings also leads to additional vortex filaments aligned in the direction of flight (and therefore perpendicular to the wings; see Figure 34(b)). Called trailing vortices, they are visualized in Figure 34(c) behind a Lockheed L-1011 Tristar using smoke generators, but can also be seen naturally thanks to the condensation of water vapour from the atmosphere at the centre of the vortices. These vortex filaments are unstable, so from long and smooth tubes they quickly transition to wavy filaments, leaving streaks in the sky that can often be seen long after the plane has disappeared from view.

Magnus effect and vorticity in sport

The dynamics of sports balls is another situation where lift is present. The lift allowing aeroplanes to fly was due to the starting vortex inducing flow circulation around the wing. Alternatively, this circulation can be generated by spinning the body moving in the fluid, and in that case it is called the Magnus effect after the 19th-century German chemist and physicist Heinrich Magnus. During ball games, the Magnus effect is used by adept players to manipulate the motion of the balls.

The Magnus effect can be illustrated using the example of tennis, as in Figure 35, where two popular techniques based on it can be used to modify the trajectory of a ball. The first one is called backspin, where the player hits the ball while slicing it downward (Figure 35(a)). This leads to a rotation that, when observing the ball moving from right to left, is in the clockwise direction. Using the same physical arguments as those explaining the lift on aeroplanes we see that backspin also leads to an upward lift force.

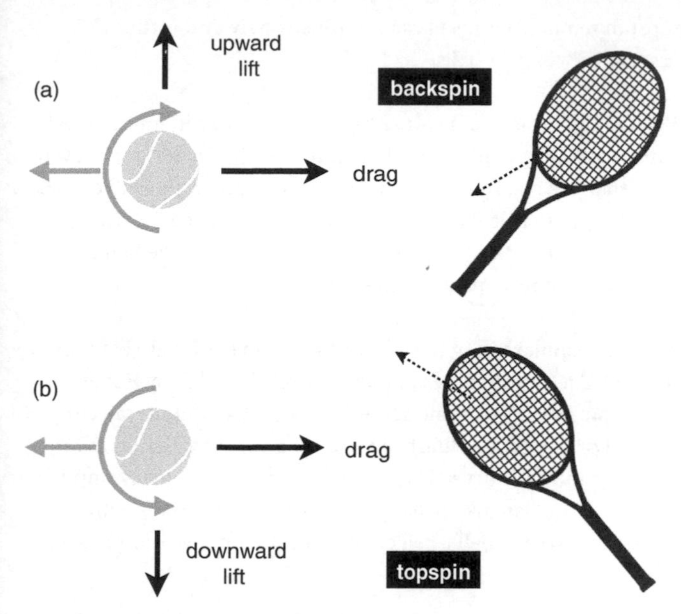

35. Lift can be induced on bodies moving through a fluid by rotating them, for example a tennis ball given spin by a racket. Two techniques can be used to induce lift. In backspin (a), the ball is given a clockwise rotation (when observing the ball moving from right to left), which induces an upward lift and prolongs the time it spends in the air. In contrast, with topspin (b) the ball rotates in the counter-clockwise direction, and the associated downward lift brings it down quickly.

As a result of backspin, balls spend more time in the air, with a bell-like curve, giving the player extra time, for example to run to the net.

The second technique is called topspin, in which the player hits the ball while slicing it upward (Figure 35(b)). In that case, the rotation of the ball is in the counter-clockwise direction. Due to the change in rotation direction as compared to backspin, the ball experiences a downward lift, which accelerates its descent. To prevent the ball from dropping too soon and touching the net, topspin requires tennis players to hit strongly and is thus associated with fast balls.

Beyond tennis, the same Magnus effect is at play in multiple ball games. In table tennis, the ball is light and thus easy to spin even by casual players. In baseball, pitchers who throw the ball can use their fingers to give it a rotation as it leaves their hands. In volleyball, during serves, players often add spin to the ball to control the time it spends in the air.

Another popular use of the Magnus effect in sports are kicks in football. An (in)famous example is a 1997 free kick by Roberto Carlos for Brazil in a game against France. The initial trajectory of the ball projected it to finish far from the goal but, thanks to strong sidespin, the ball curved back into the net. A different example in football is the Panenka penalty kick, where a slow kick with backspin gives the ball a bell curve trajectory above the goalkeeper.

Chapter 7
Instabilities

A clear difference exists between the canonical, and relatively well-defined, flows considered so far in the book and the complicated fluid motion experienced in our daily life. The key to understanding the formation of these complex fluid motions is through the study of flow instabilities. A fluid flow is said to be unstable if a small perturbation to it does not decay to zero but instead becomes amplified with time. Since real life is full of unavoidable perturbations, these instabilities are the building blocks for countless intricate flow patterns. In this chapter, we will look at a physical description of fundamental instability mechanisms in fluid mechanics.

Gravitational instability: Rayleigh–Taylor

It is instructive to start with a simple example, the most intuitive one being the Rayleigh–Taylor instability. Consider two different fluids at rest. Invariably they will have different mass densities. For example, olive oil in salad dressing is approximately 10% lighter than the vinegar, so dressing bought at the store has the oil at the top and the vinegar at the bottom.

This intuitive light/heavy conformation is actually the result of a fluid mechanics instability. Lord Rayleigh was the first to show formally that the situation where the heavy fluid is located below the light one is stable, whereas the reverse situation with the heavy fluid on top is unstable. This instability is illustrated in Figure 36(b) using a heavy dye released on top of water in a glass. The initial configuration with light water at the bottom is unstable and the heavy dye is seen to progressively penetrate into the water along elongated threads with characteristic protrusions.

The physical argument for this instability is a net release of gravitational potential energy (GPE) in all situations where a parcel of the heavy fluid is located above a lighter one. Consider the flat horizontal interface of the hypothetical setup where all the lighter fluid is below the heavy one (see Figure 36(a)). Since that

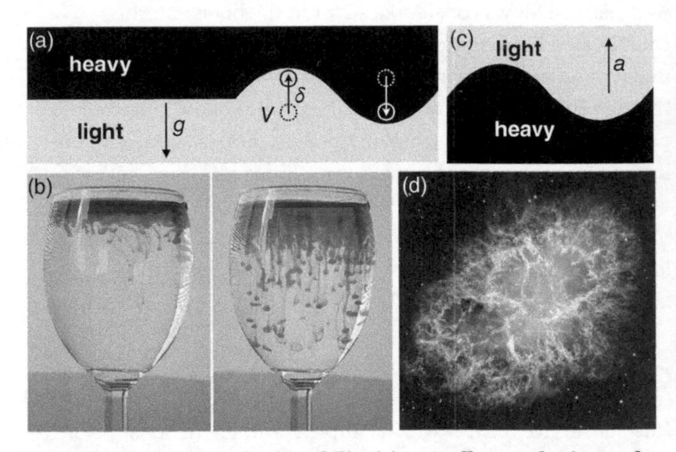

36. In the Rayleigh–Taylor instability (a), a small perturbation to the interface between a heavy fluid located above a lighter one releases gravitational potential energy and is therefore amplified, leading to an instability; this is illustrated using two small volumes V displaced by a magnitude δ in each fluid. An experimental illustration of this instability is achieved by depositing a heavy dye in a glass of water (b). A similar instability arises when a fluid–fluid interface is accelerated in the direction from the heavy to the lighter fluid (c). This is illustrated by the patterns in the Crab Nebula, where the cold and heavy remains of a star's supernova explosion are accelerated inward towards lighter plasma, leading to characteristic Rayleigh–Taylor protrusions (d).

situation is unstable, it is of course just a thought experiment. Now imagine perturbing this interface in some arbitrary wave-like pattern. Doing so allows some of the top fluid to move down and some of the bottom fluid to move up. From the point of view of GPE, this is an exchange: some light fluid replaces heavy fluid and vice versa, and therefore the total energy of the system decreases.

To make this quantitative, consider two small volumes V in the fluid, one in the bottom fluid and one in the top fluid, both near the interface (see Figure 36(a)). The GPE of each volume is given by mgh where h is the height of their centre of mass and $m = \rho V$ is the mass of the fluid. The total GPE before the perturbation of the interface, GPE_{old}, is then

$$GPE_{\text{old}} = \rho_{\text{heavy}} Vgh_{\text{heavy}} + \rho_{\text{light}} Vgh_{\text{light}} \qquad (26)$$

We denote by δ the size of the vertical perturbation of the interface, and therefore of the vertical displacement of each small fluid volume. The heavy fluid volume goes down by an amount δ so its new GPE is $\rho_{\text{heavy}} Vg(h_{\text{heavy}} - \delta)$. Similarly, the light fluid goes up by δ so its new GPE is $\rho_{\text{light}} Vg(h_{\text{light}} + \delta)$. As a consequence, the total GPE after motion of the interface is

$$GPE_{\text{new}} = \rho_{\text{heavy}} Vg(h_{\text{heavy}} - \delta) + \rho_{\text{light}} Vg(h_{\text{light}} + \delta), \qquad (27)$$

so the change in GPE due to the displacement of the interface, ΔGPE, is

$$\Delta GPE = GPE_{\text{new}} - GPE_{\text{old}} = Vg\delta(\rho_{\text{light}} - \rho_{\text{heavy}}). \qquad (28)$$

Since $\rho_{\text{light}} < \rho_{\text{heavy}}$, we have $\Delta GPE < 0$ and the motion of the interface always leads to a net release of gravitational potential energy. It is therefore favourable for the system to encourage this exchange, and it continues to be the case until all light fluid has moved above all the heavy fluid, at which point the minimum energy has been reached and the system is stable. This is the final state of the salad dressing.

Rayleigh's work considered the case where the overturning of the fluid is driven by gravity. That instability also bears the name of the 20th-century British physicist Geoffrey Ingram Taylor, universally hailed as the founder of modern fluid mechanics and, arguably, the most influential fluid mechanician of all time. Celebrated for his work combining both experiments and theory and for his creativity, Taylor is also revered for being not just a physicist who dabbled in fluid mechanics, but somebody who wholeheartedly embraced the field in its totality, and as a result spawned multiple generations of aspiring fluid mechanicians.

Taylor realized that, beyond gravity, a similar instability could take place in the case where fluids with different densities are accelerated. Indeed, the strength of the gravitational field g is nothing but the acceleration due to gravity, and a body in free fall has an acceleration of magnitude exactly equal to g towards the Earth. Taylor realized that even the stable configuration of light fluid sitting on top of heavy fluid could be rendered unstable if the two fluids were subject to an upward acceleration a of magnitude larger than g (see the schematic illustration in Figure 36(c)). A liquid–liquid surface is then unstable when it experiences a net acceleration in the direction from the heavier to the lighter fluid.

This version of instability is why we shake our bottle of salad dressing: the acceleration provided to the stable oil–vinegar configuration allows us to mix them. The famous Crab Nebula is another example, shown in Figure 36(d). The cold and heavy remains of a star's explosion into a supernova accelerate inward towards lighter plasma, resulting in patterns characteristic of the Rayleigh–Taylor instability.

Stability theory

The Rayleigh–Taylor instability, which we illustrated using experiments and physical arguments, can be studied from a more theoretical perspective. A well-established mathematical framework, called stability theory, has been developed to hunt for instabilities—not just in fluid mechanics but in many topics of physical and life sciences. It belongs to a branch of mathematics called dynamical systems, which aims to study the solutions to differential equations and the possible transitions in the system from one solution to another.

Stability theory works approximately in the following way. We are equipped with the equations that describe the phenomenon we are trying to understand. In the case of fluid mechanics, we have the Navier–Stokes equations. The first step is to consider a known solution of these equations. That solution might not be easy to realize in the lab but it is a mathematical solution that is achievable under the right conditions. It is called a base state and in fluid mechanics we often refer to it as a base flow. The latter is in general a very simple flow, e.g. the two quiescent fluids sitting on top of one another in Figure 36. Sometimes, the base flow is complex and it has to be found computationally.

The second step is to perturb the base flow by a small amount and see what happens. If all perturbations decay in time, the flow is stable. If some perturbations grow, then we determine which is the one that grows the fastest. The rationale behind this is based on experimental observations: if a particular perturbation to the flow grows faster than all the other ones, it will eventually take over the whole fluid dynamics and will therefore be the one measured experimentally.

However, how can we figure out what part of a perturbation is the most important? After all, a system can be perturbed in an infinite number of different ways. Drawing from over 300 years of research, we usually know a natural way to project mathematically any perturbations on a set of so-called natural modes for the system. Although the theory underpinning this result is complicated, it makes intuitive sense. Fluid systems are analogous to musical instruments, which are characterized by intrinsic modes, e.g. the vibrations of the strings on a guitar. In most cases, the natural modes are sinusoidal waves. The fundamental reason is due to celebrated work from Fourier, who showed how any arbitrary mathematical function can be written as a superposition of so-called Fourier modes, all taking the shape of sinusoidal functions. A crucial parameter of these modes is their wavelength, λ, which is their spatial periodicity; in many experiments on flow stability, the first thing that we measure is the wavelength of the instability.

Equipped with these modes, the final step is to then identify which one grows the fastest, which is called a modal growth analysis. The final result is a prediction for the most unstable wavelength (i.e. the wavelength of the most unstable mode) and the rate at which the mode grows. Applying this procedure to the Rayleigh–Taylor instability, we find that any wave-like mode of surface deformation is unstable. For an infinite fluid interface with no surface tension, we obtain the result that the larger the wavelength of the wave, the faster the growth of the interface distortion between the two fluids.

Historically, hydrodynamic instabilities have often been named after the people who discovered them and elucidated their physical mechanism. This is the case for the Rayleigh–Taylor instability, with first Rayleigh and then Taylor providing the physical and mathematical understandings we have today. Taylor, whose name is associated with three different instabilities in this chapter, is famous for his 1922 paper on the centrifugal instability (see below), which reported the first quantitative agreement between stability theory and experiments. Since then, stability theory has been one of the most popular and successful avenues of research in fluid mechanics due to its ability to predict and explain experimental results.

Capillary instability: Rayleigh–Plateau

The second instability in this chapter originates from surface tension, which we saw in Chapter 1 is an energy associated with the presence of fluid interfaces. The fact that a liquid has surface energy leads to the famous Rayleigh–Plateau instability wherein long fluid threads cannot remain elongated but break up into small droplets. The physical rationale behind this instability is remarkably simple: an elongated liquid thread of given volume can decrease its total surface area, and therefore its surface energy, if it changes form and separates into multiple droplets. So we can now start blaming surface tension for our dripping taps (Figure 37(a)).

A simple mathematical approach helps to understand this instability. Consider an elongated cylinder of liquid of length L and radius R as in Figure 37(b). The total volume of fluid is $V = \pi R^2 L$ and if it is long enough its surface area is approximately $A \approx 2\pi RL \approx 2V/R$. Now let us imagine cutting this cylinder into N identical droplets, each of radius a. Conservation of volume imposes the relationship $N \times 4\pi a^3/3 = V$ so the number of droplets is $N = 3V/4\pi a^3$. The total surface area, $\tilde{A}$, of the N droplets is equal to $\tilde{A} = N \times 4\pi a^2 = 3V/a$. The ratio between the total surface area of the droplets to that of the original fluid thread is given by $\tilde{A}/A = 3R/2a$. Hence, the droplets are predicted to have a surface area smaller than that of the cylinder if $a > 1.5R$ and it is thus energetically favourable for an elongated cylinder of fluid to break up into droplets if they are large enough.

The associated instability, sometimes referred to as surface-tension-driven breakup, or capillary instability, is associated with the work of Joseph Plateau. A 19th-century Belgian physicist and former child prodigy, Plateau's work on surface tension and droplets laid the groundwork for much of the

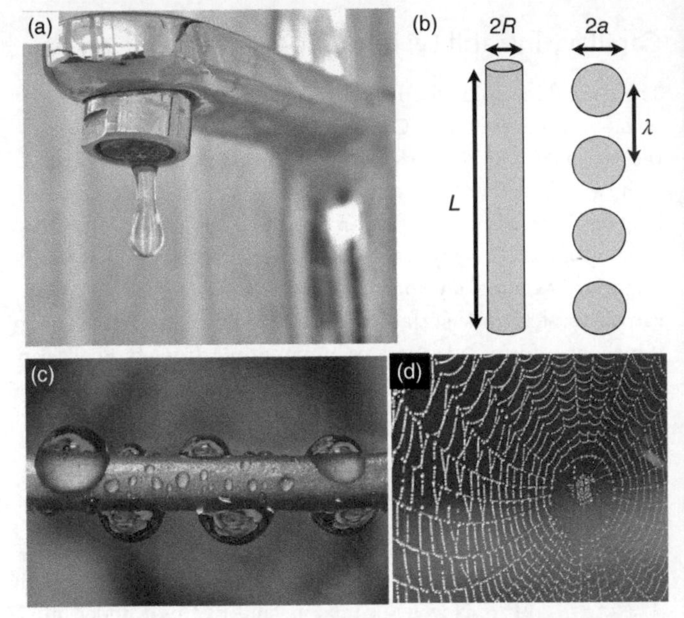

37. In the capillary (or Rayleigh–Plateau) instability, a liquid thread spontaneously breaks up into small droplets, as seen in many everyday examples such as dripping taps (a). Physically, an elongated cylindrical thread of length *L* and radius *R* can decrease its total surface area, and hence its surface energy, by breaking up into droplets of radius *a*, provided they are large enough (b). The capillary instability also occurs in nature, for example for water on plant stems (c) or dew on spider webs (d).

modern science of thin films and foams. He was also a pioneer in the development of moving images and animated cartoons.

Beyond the physical picture outlined above, Plateau realized that we need a bit more than a final state of lower energy. For the instability to happen spontaneously, surface energy needs to be released as soon as the interface is perturbed. Considering infinitesimal sinusoidal perturbations with wavelength λ of the liquid thread from Figure 37(b), Plateau showed rigorously that as long as $\lambda > 2\pi R$, these small perturbations will decrease the area of the interface, freeing up some surface energy to accelerate the liquid and allowing the process to continue until pinch-off and the creation of droplets.

What does Plateau's analysis mean for the final sizes of the resulting droplets? When a liquid thread is unstable with wavelength λ and every periodic contraction of the thread collapses into a single droplet, conservation of volume implies that the radius a of each droplet is given by $4\pi a^3/3 = \lambda\pi R^2$ and thus $a = (3\lambda R^2/4)^{1/3}$. Plateau's criterion for the spontaneous instability, $\lambda > 2\pi R$, leads then to a minimum size $a > R(3\pi/2)^{1/3}$, which is approximately $a > 1.68R$. This criterion is slightly more stringent than the argument based solely on the consideration of the surface energy of the final state ($a > 1.5R$).

How does this compare with the size of droplets seen in experiments? The energetic arguments suggest that the largest decrease in surface energy occurs for a (or λ) as large as possible above Plateau's threshold. Indeed for a given volume of liquid, a mathematical theorem states that a sphere of the same volume is the shape with the smallest possible surface area. So starting from any fluid mass, the state of lowest surface energy will be the one with all the fluid gathered into a single spherical drop.

However, this is not what is observed experimentally. Liquid threads breaking up into droplets do so at a very specific wavelength, in a manner that is well reproducible experimentally. The fundamental reason is that although surface tension is the driving force for the instability, what is observed in experiments is not the mode with the least surface energy, but the mode that grows the fastest. Of all the possible ways in which the interface can be perturbed, one grows faster than any other and is the one seen experimentally. That mode is set by a competition between the driving and the dynamics.

Why exactly intermediate wavelengths are preferable was elucidated by Rayleigh, who carried out experiments on liquid threads and then showed theoretically that λ cannot be too large (despite the fact that the larger the value of λ, the smaller the final surface energy). Indeed, if λ was large, the instability would involve creating periodically very long threads of liquid. These elongated threads would then need to contract in order to make drops. Since the fluid starts from rest, each thread needs a certain time to accelerate from rest, and the longer the thread the longer it takes to set it in motion. The most unstable wavelength seen in experiments is therefore a compromise: it needs to balance the requirement of releasing a sufficient amount of surface energy (λ large enough) with having the instability occurring sufficiently fast (λ small enough). Rayleigh's calculation showed that the wavelength growing the fastest was $\lambda \approx 9R$, well above the threshold based solely on energetics ($\lambda > 2\pi R$) and in excellent agreement with experimental observations. Because of both Plateau and Rayleigh's important contributions, this phenomenon is now known as the Rayleigh–Plateau instability.

While the classical version of the instability is for a free-standing liquid column, it takes many other forms in our daily lives, and as such has been a source of study for over 200 years. For example, instead of a free-standing liquid, a thin film deposited on a surface can also be unstable. Examples are films of liquid located on the ceiling, e.g. due to humidity in a bathroom or to water vapour in a kitchen. Surface tension would prefer the film to remain flat but gravity would like the fluid to fall down. As a compromise, the film spontaneously breaks up into an array of small droplets. Another example is liquid deposited on a rigid fibre. The same physical principle as the classical Rayleigh–Plateau instability (a decrease of surface energy) predicts that the liquid will break up into droplets. This phenomenon is familiar in nature, seen for example for water drops on plant stems in Figure 37(c) or dew droplets on spider webs in Figure 37(d).

Centrifugal instability: Taylor–Couette

Returning to fluids in motion, another instability occurs for rotating flows, called the centrifugal instability. It is a historically important example because it is the first time a quantitative understanding of a Navier–Stokes instability was shown to be in full agreement with experiments.

The simplest setup consists of a fluid located in the space between two rotating coaxial cylinders, as illustrated in Figure 38 ((a) side view; (b) view from above). The geometry is reminiscent of that of a shear flow, originally studied by Couette (see Chapter 2), but adapted to the circular geometry by Taylor, and it is now known as a Taylor–Couette cell. The same geometry is used in many rheometers to measure the mechanical properties of complex fluids.

As the cylinders turn in the Taylor–Couette cell, the fluid in contact with them also rotates. Consider now what happens when we change the angular rotation of the inner cylinder (keeping the outer at a fixed speed). When the inner cylinder rotates slower than some critical value, the flow is stable and all parcels of fluid just rotate around. However, if the inner cylinder rotates faster than that threshold, the flow becomes unstable and it transitions to a new dynamical state exhibiting three-dimensional doughnut-like vortices oriented perpendicular to the axis of rotation (Figure 38(c)–(d)).

The fact that a rotating flow could be unstable was anticipated by Lord Rayleigh. Back in 1888, he derived a criterion for such an instability to occur in the case where the role of viscosity can be neglected. To understand the origin of the instability, von Kármán proposed an explanation of Rayleigh's criterion based on an analysis of the forces experienced by the fluid. Consider a small parcel of fluid located at a distance r from the axis of rotation and moving with speed u. If Ω is the local angular speed of the rotating

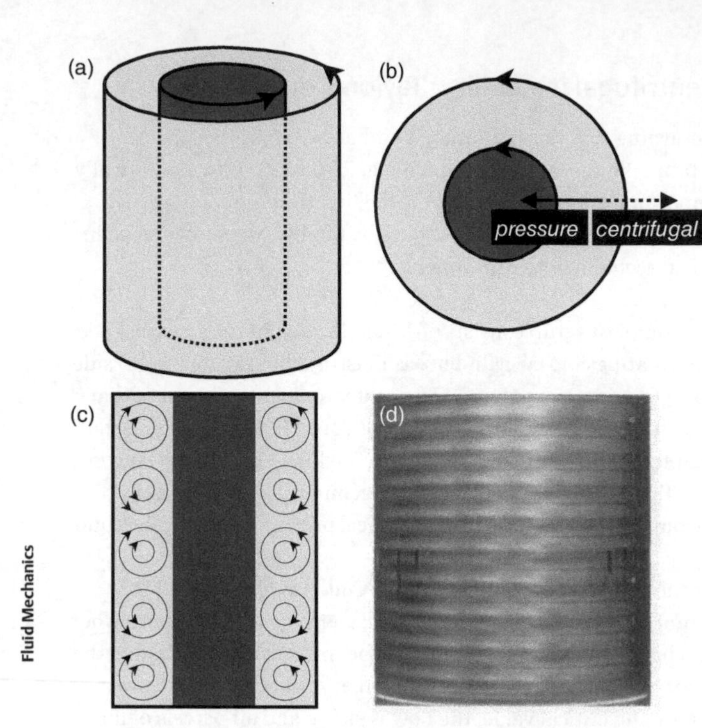

38. The centrifugal instability in a Taylor–Couette geometry can appear for a fluid located in the space between two rotating cylinders ((a) side view; (b) view from above). At each point in the fluid, the outward centrifugal force balances the inward resistance due to the pressure in the fluid (b). If the inner cylinder rotates too fast, the flow is unstable and it transitions to a regime where vortices of alternating directions are created between the cylinders ((c) sketch; (d) experiments). A further increase of the inner cylinder velocity leads to even more complex flow and, ultimately, turbulence.

fluid, we have $u = \Omega r$. It is a classical result of physics that a body in rotation is subject to a centrifugal force directed radially outwards and of magnitude $F_{\text{out}} = mr\Omega^2$.

However, before the instability, the fluid is not seen to move outwards, so what acts against the centrifugal force? Since the fluid rotates around the axis of the cylinders, an inward fluid force, F_{in}, must be balancing the outward centrifugal force. This fluid force is in fact due to the distribution of pressure, which adjusts throughout so as to exactly balance the centrifugal force. This is similar to the hydrostatic pressure generated in fluid subject to gravity (see Chapter 1), where the gravitational force balances the force due to pressure in the fluid. In the case of the rotating fluid, there is similarly at every point in the fluid an equilibrium between the outward centrifugal force (dotted arrow in Figure 38(b)) and the inner force due to the pressure (solid arrow), i.e. $F_{in} = F_{out}$.

To understand the instability, we then have to perturb the moving fluid. However, because the fluid rotates, it possesses angular momentum and, in the absence of viscous friction, any perturbation can only be made in a way that conserves its momentum. We saw already in Chapter 6 via Helmholtz's vortex laws (and Kelvin's circulation theorem) some consequences of the conservation of angular momentum for flows. The angular momentum for a small parcel of fluid of mass m is equal to $L = mru = mr^2\Omega$ and since it is conserved, we can express the magnitude of the two balanced forces discussed above as $F_{in} = F_{out} = mr\Omega^2 = L^2/(mr^3)$.

To analyse the fate of a perturbation, let us consider a small parcel of fluid initially located at a distance r_1 from the axis of rotation and with angular momentum L_1. Suppose that fluid was then displaced to new location $r_2 > r_1$, thus taking the place of a parcel of fluid that had angular momentum L_2. Since the fluid keeps its angular momentum with it, the outward centrifugal force exerted on the parcel of fluid that moved to r_2 with its momentum L_1 now has magnitude $F_{out}^{new} = L_1^2/mr_2^3$. However, except for this perturbation, the surrounding fluid has not changed, so the distribution of pressure remains identical. The radially inward pressure force is thus still of the magnitude it was before the new fluid parcel came in, i.e. $F_{in}^{new} = L_2^2/mr_2^3$, where L_2 is the angular momentum of the fluid that has been displaced.

To decide if the flow is stable, we need to compare the magnitude of the two forces exerted on the parcel of fluid that has been displaced. If $F_{\text{out}}^{\text{new}} < F_{\text{in}}^{\text{new}}$, i.e. $L_1^2 < L_2^2$, the force from the pressure wins and a net inward force brings the fluid back to where it started. The flow is therefore stable. In contrast, when $F_{\text{out}}^{\text{new}} > F_{\text{in}}^{\text{new}}$, i.e. $L_1^2 > L_2^2$, the outward force wins and the parcel of fluid is pushed further away from the axis of rotation. In that case, the initial perturbation is amplified and the flow is unstable.

This result is the stability criterion originally derived by Rayleigh. The fluid flow is subject to the centrifugal instability if the magnitude of the angular momentum decreases away from the axis of rotation at one point in the fluid. Given the relation between circulation and angular momentum, this is similar to saying that, for the instability to take place, the circulation has to decrease radially outwards somewhere in the fluid. Note that an alternative physical argument can be used to predict the instability by considering the total kinetic energy in the displaced parcels of fluids. The exchange of fluid parcels turns out to also be energetically favourable as soon as $L_1^2 > L_2^2$.

The stability analysis of Rayleigh ignored the friction in the fluid (i.e. the viscosity) in order to make use of the conservation of angular momentum. In a landmark 1923 paper, Taylor set out to study the viscous version of this problem theoretically and experimentally. He first carried out a full stability calculation for the flow and was able to solve the resulting mathematical equations exactly—a heroic task since this was well before personal computers could solve differential equations. His theoretical results derived the region in parameter space where the flow was predicted to be stable or unstable, and they were in perfect agreement with experiments. This study directly comparing a full stability theory with experiments was the first of its kind and it has remained the benchmark in the field for almost 100 years.

Past the stability point, the rotational flow transitions to a new state. Fluid parcels are no longer just rotating with the cylinder but they display three-dimensional motion. Just after the instability, the flow takes the form of alternating doughnut-like eddies aligned perpendicular to the axis of rotation of the cylinder. Now called Taylor vortices, they are illustrated schematically in Figure 38(c) and shown experimentally in Figure 38(d). If the angular speed of the inner cylinder is increased even further, the flow undergoes a cascade of additional instabilities and is now used as a canonical case study of the transition to turbulence.

Viscous fingering instability: Saffman–Taylor

We next consider a famous instability occurring between two fluids of different viscosities. The prototypical situation consists of one fluid pushed against another one under confinement (Figure 39(a)). When the displacing fluid has a lower viscosity than the one it replaces, the moving interface between the two fluids is unstable to small perturbations (Figure 39(b)) and develops interpenetrating protrusions of each fluid moving into the other one, resulting in a characteristic finger-like pattern (Figure 39(c)). This instability is known as the viscous fingering instability, and it turns out to be important in industry (Figure 39(d)).

To understand the physical mechanism for the instability, we have to examine the dynamics of a moving interface between two fluids. For the instability to take place, the fluids need to be geometrically confined to ensure that the flow is dominated by viscous friction with its environment. This setup is generically referred to as a porous medium and classical examples of porous media include rocks, cement, wood, and bones.

The key feature of all flows dominated by viscous friction is something we already saw in Chapter 3, namely the

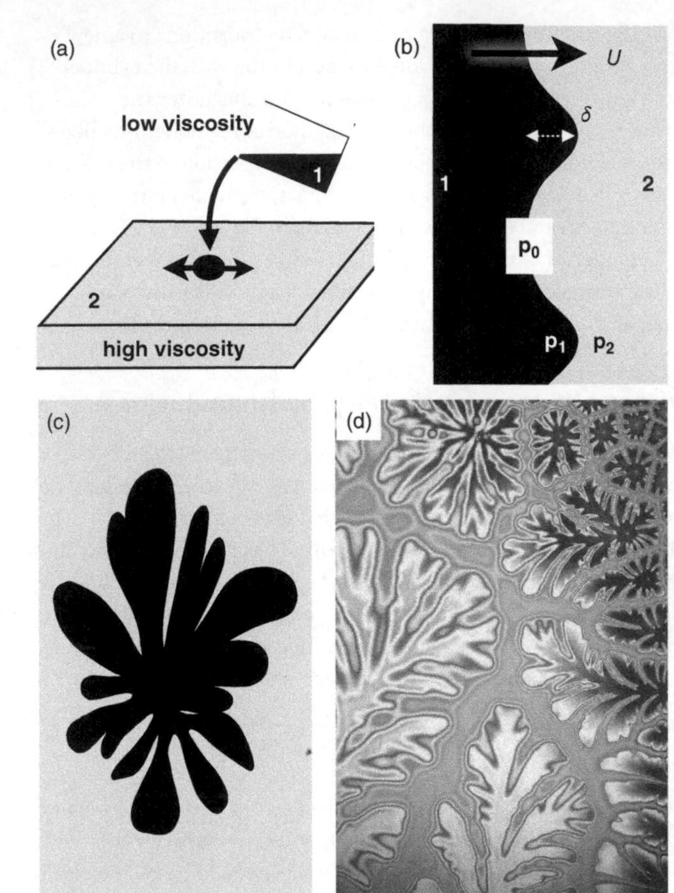

39. The viscous fingering (or Saffman–Taylor) instability arises when a low-viscosity fluid (here, denoted by #1) is used to displace a fluid with a higher viscosity (#2) in a domain where the flow is dominated by viscous friction (a). In that case, a small perturbation to the interface between the two moving fluids is amplified, which can be rationalized by comparing the values of the fluid pressures p_1 and p_2 on either side of the perturbation of size δ (b). The instability results in a characteristic shape of elongated protrusions of each fluid into the other (c). The viscous fingering pattern is illustrated experimentally using a colloidal suspension displaced by air (d).

Hagen–Poiseuille law, Equation (4). That formula predicted the flow rate in a pipe as a function of the applied pressure. A similar formula is applicable to all flows dominated by the effects of viscosity. In all these cases, over a length L, we always have a pressure drop ΔP that is proportional to the average fluid velocity U and viscosity μ in the form

$$\frac{\Delta P}{L} = \frac{\mu U}{\kappa}, \tag{29}$$

where κ has units of square of a length and is called the permeability of the material. This equation is known as Darcy's equation for flow in porous media. The permeability κ is purely a function of the geometry of the domain through which the fluid moves. We know its exact value for many geometries, for example the straight pipe from Chapter 3 (it is equal to $D^2/32$ when D is the diameter of the pipe; see Equation (4)), but also for flow between two rigid surfaces ($\kappa = H^2/12$ where H is the distance between the surfaces), flow arounds arrays of obstacles, etc. The relationship between the geometry of the material and its permeability is therefore known.

The result in Equation (29) is the key to understanding the fingering instability. Consider the moving interface illustrated in Figure 39(b), where the fluid of viscosity μ_1 is advancing towards that of viscosity μ_2. The interface is straight and perpendicular to the direction in which it moves with speed U. Now imagine a small perturbation, of size δ, to the location of the interface. When $\delta > 0$, the perturbation is in the direction of motion, i.e. from fluid 1 to fluid 2 (this case is illustrated in Figure 39(b)) and in the other direction if $\delta < 0$.

To analyse the fate of this perturbation, we apply Darcy's equation to both fluids and compute the change in pressure induced over the small distance δ. Let us call the pressure at the location of where the interface would be if there were no perturbation p_0. The formula in Equation (29) says that the pressure decreases in the direction of the fluid motion, so when applied to fluid 1 we obtain a change in the local interfacial pressure to a new value p_1 given by $(p_0 - p_1)/\delta = \mu_1 U/\kappa$ (see Figure 39(b)). When $\delta > 0$, the pressure therefore decreases slightly near the interface on the side of fluid 1, or increases slightly when the perturbation is the other way ($\delta < 0$). The same argument can be applied to fluid 2, so the interfacial pressure p_2 on the side of fluid 2 satisfies $(p_0 - p_2)/\delta = \mu_2 U/\kappa$. These two equations give us access to the pressures on either side of the moving interface and the difference in pressure is

$$p_1 - p_2 = \frac{(\mu_2 - \mu_1)U\delta}{\kappa}. \tag{30}$$

The result in Equation (30) is the key to predicting the evolution of the perturbation. Given that $U > 0$, let us first consider the case with $\delta > 0$. If $\mu_2 < \mu_1$ (i.e. the displacing fluid is the more viscous one), then we see that $p_1 < p_2$ and the pressure from fluid 2 is larger; a net force is thus applied to the interface that stabilizes the perturbation. In contrast, when the less viscous fluid is moving towards the more viscous one, i.e. $\mu_2 > \mu_1$, we obtain that $p_1 > p_2$; in that case, a net pressure force is exerted on the interface and the initial perturbation is amplified, leading to an instability. The same result holds when the perturbation is directed towards fluid 1 ($\delta < 0$). Here, $\mu_2 > \mu_1$ leads to $p_1 < p_2$ so a net force is applied in the direction of the perturbation, again indicating an instability.

This analysis therefore predicts the occurrence of an instability if the interface moves from the less to the more viscous fluid (i.e. if the product $(\mu_2 - \mu_1)U$ is positive). This is the fundamental condition for the fingering instability to appear. Under that condition, protrusions in both directions are amplified, and result in the characteristic patterns of interlocking fingers illustrated in Figure 39(c). An experimental illustration of the instability is shown in Figure 39(d), obtained with air displacing a colloidal suspension.

This instability is also known as the Saffman–Taylor instability. Philip Saffman was a British-American applied mathematician who studied in Cambridge. In collaboration with Taylor, he ran experiments demonstrating the instability, which had been predicted a few years earlier, carried out the full stability analysis, and further focused on the shape of fingers. After moving to the US, he became famous for many other contributions to fluid mechanics, including vortex dynamics and flows in biological membranes.

Viscous fingering is important in many industrial processes, most famously in oil recovery. When crude oil does not flow out of an underground reservoir on its own, different techniques can be used to extract it, called enhanced oil recovery. For example, gases can be injected (such as CO_2), but also steam or water. The issue is that crude oil can be very viscous, hence the conditions are often such that a less viscous fluid is used to push against a more viscous one. As a result, the Saffman–Taylor instability occurs and the recovery is very inefficient. One remedy is to mix polymers with water to increase its viscosity sufficiently so as to prevent the instability. Other industrial situations where viscous fingering plays an important role include hydrology, filtration, and CO_2 sequestration.

Shear layer instability: Kelvin–Helmholtz

Our final case study in the world of flow instabilities arises when two fluid streams moving at different speeds are put in contact with one another. This setup is known as a shear layer, a region where a large jump in fluid velocity occurs on a small length scale, and it is illustrated in Figure 40(a). As a result of this difference in speeds, the interface between the two fluids starts to develop a wave-like motion. These waves are further amplified downstream, rolling up like surface waves in the ocean and resulting in the creation of large vortices separating the two fluids.

This instability was first discovered by Helmholtz and Kelvin independently from one another within a span of three years, and

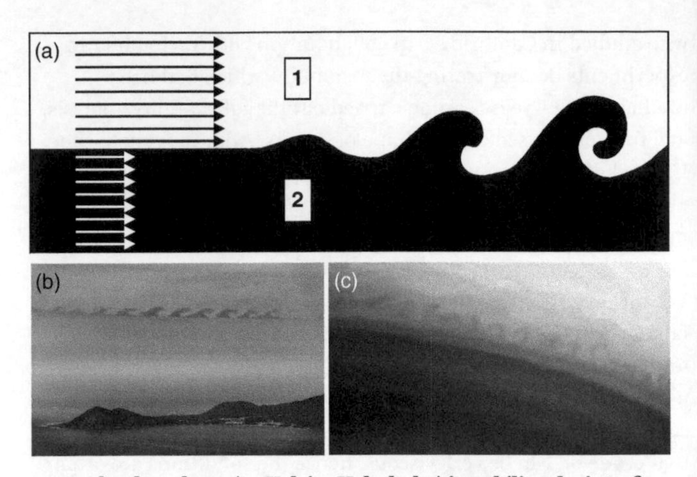

40. In the shear layer (or Kelvin–Helmholtz) instability, the interface between two fluid streams moving with different speeds is unstable so any small surface perturbation is further amplified and the interface eventually rolls up like ocean waves (a). The instability can often be seen in the atmosphere with the aid of clouds (b) and has also been reported to occur in astrophysics, for example in the gas atmosphere of Saturn as photographed by NASA's Cassini probe (c).

is now known as the Kelvin–Helmholtz instability. It can be rationalized using a physical principle first seen in Chapter 5 for flows at high Reynolds numbers. For a perfect fluid with no swirling motion, we saw that conservation of energy in the fluid led to Bernoulli's equation for the pressure p, density ρ, and speed U stating that the combination $p + \rho U^2/2$ is constant (Equation (20)). The consequence of that equation is that, for such flows, when the speed increases, the pressure decreases; we already used the same idea in Chapter 6 to elucidate the lift generated on aeroplane wings.

To explain why a shear layer is unstable, consider the flat interface between the two fluids and let us imagine perturbing it as shown in Figure 40(a). The initial perturbation of the interface is directed towards fluid 1 and therefore it decreases slightly the total area available for fluid 1 to flow. Recall from Chapter 3 that, due to conservation of mass, the flow rate, which is equal to the product of speed with area, must be constant for an incompressible flow. If the area decreases, the speed of fluid 1 therefore needs to increase slightly to compensate for it. Similarly, since the perturbation is directed away from fluid 2, the area available for its flow increases, so for the same reason fluid 2 slows down slightly.

We can then use Bernoulli's equation to estimate the change in pressure in each fluid resulting from the perturbation. Let us denote by p_0 the pressure in the fluids near the interface before any perturbation and by p_1 and p_2 the new fluid pressures near the interface inside fluids 1 and 2, respectively. In fluid 1, the speed of the fluid increases in magnitude and therefore, from Bernoulli, pressure necessarily decreases, so $p_1 < p_0$. Similarly in fluid 2, the fluid speed decreases and thus the pressure has to increase, i.e. $p_2 > p_0$. The initial perturbation of the interface shown in Figure 40(a) therefore leads to changes in pressure on both side of the interface with $p_2 > p_1$. As a result, a net pressure force is induced in the direction of the perturbation, which amplifies it and leads to an instability. Of course, if instead the perturbation were initially directed towards fluid 2, a similar analysis would lead to $p_1 > p_2$, again indicating an instability.

This Kelvin–Helmholtz instability occurs in situations where two fluids flowing at different speeds are brought together, and it impacts a wide range of processes in industry and nature. For example, it takes place in the ocean in all instances where underwater currents flowing at different speeds merge. We can also routinely see the instability in the atmosphere, as illustrated in Figure 40(b) where the rolling up of the vortices is visualized with the aid of clouds. The same instability has been seen on other planets, as illustrated in Figure 40(c) with the gas atmosphere of Saturn as visualized by NASA's Cassini probe; even though the flow is in the turbulent regime in that case, the phenomenology of the instability remains valid.

In industry, the Kelvin–Helmholtz instability is an important mechanism to generate mixing between different fluids. This can be intuitively understood when examining the sketch in Figure 40(a). During the development of the instability, the interface between the two fluids rolls up so the interfacial area increases; as the instability develops downstream of the initial contact, the two fluids further intertwine. The resulting mixing can have major consequences. For example, if the two fluids react chemically to produce a third product, then the instability will promote the chemical reaction. This is what happens inside your car engine where the instability facilitates the mixing between different reactants and helps combustion. If the two fluids are at different temperatures then the instability will help the homogenization of the temperature. In the ocean, it also contributes to the mixing of nutrients for biological organisms.

Chapter 8
Researching fluids and flows

Today, active research in fluid mechanics continues in a multitude of topics; I highlight some notable ones in this chapter.

Environmental flows

Humankind is facing an indisputable crisis in the form of climate change. Fluid mechanics plays an active role in helping us understand the active dynamics of our changing environment, from large-scale atmospheric flows to transport in oceans and the melting of glaciers at the poles. The flows of rivers, lakes, and other reservoirs also contribute to the evolution of our climate. One important aspect is stratification, i.e. the fact that a fluid can have a density that depends on depth, for example due to salt in the ocean or temperature differences in the atmosphere, which in turn affects many of the topics summarized in the previous chapters.

Fluid mechanics of energy

A related environmental topic at the heart of building a greener world is the fluid mechanics of energy. The design of wind farms requires an understanding of the unsteady wake flows behind wind turbines and their collective interactions. Fluid mechanics can inform the work of architects and help them exploit natural ventilation flows arising from wind and temperature differences to design green buildings. On small scales, the efficient extraction of biofuels from algae is affected by bioconvection instabilities resulting from a coupling between the cells and their fluid environment. At the other end of the energy budget spectrum, many aspects of pollution transport and remediation involve fluid mechanics, for example flows in porous media during carbon sequestration, the motion of microplastics in the ocean, and the transport of contaminants in urban environments.

Complex fluids and materials

Numerous everyday materials are neither true fluids nor true solids. A branch of fluid mechanics specializes in such complex fluids, for which the relationship between flows and stresses is non-linear and depends on the history of the flow. These are called 'non-Newtonian' fluids, in contrast with the Newtonian behaviour summarized in Chapter 2. Complex fluids fill our daily lives, from creams and gels to blood and mucus. They are also ubiquitous in our environment (for example, granular materials and sediments) and play an important role in industry, from paint and suspensions to liquid crystals and molten glass. Exciting new research is being carried out to predict the behaviour of complex fluids and to design new soft materials with tailored mechanical properties.

Flows on small scales

At the end of the 20th century, new experimental methods have enabled us to probe fluid motion down to very small length scales (nanometres). This, in turn, has allowed fluid dynamicists to discover new phenomena, from an apparent violation of the no-slip boundary condition on hydrophobic surfaces and low hydrodynamic friction in carbon nanotubes to the intricate coupling between electric fields and ions in electrolytes (e.g. water). New methods to control small suspended objects (optical tweezers) also mean we can now study in detail the hydrodynamics of fluctuating suspensions.

Biological flows

As the science of transport, fluid mechanics impacts countless aspects of biological life. From the flight of birds and insects to the swimming of fish and dolphins, locomotion at high Reynolds numbers inspires engineers to look for efficient biomimetic designs. At low Reynolds numbers, the motility of bacteria and spermatozoa powered by flagella provides us with model systems to understand the emergence of coherent motion from stochastic biological activity. Vascular transport is another topic where fluid mechanics plays a dominant role; for animals, this consists primarily in blood flow in arteries and veins along with the transport of other biological fluids such as lymph and cerebrospinal fluid. Vascular plants also rely on fluid mechanics for the transport of water up from the roots and nutrients down from the leaves. The recent Covid-19 pandemic has also shown that fluid mechanics plays an important role in the transmission of infectious diseases, in particular for airborne transmission via coughing-induced sprays (Figure 41(a)).

The holy grails of fluid mechanics

Fluid mechanics still contains, at its heart, two fundamental mysteries. The first one is turbulence. Legend has it that in a 1932 presentation to the British Association for the Advancement of Science, Horace Lamb said: 'I am an old man now, and when I die and go to Heaven there are two matters on which I hope for enlightenment. One is quantum electrodynamics, and the other is the turbulent motion of fluids. And about the former I am really rather optimistic.'

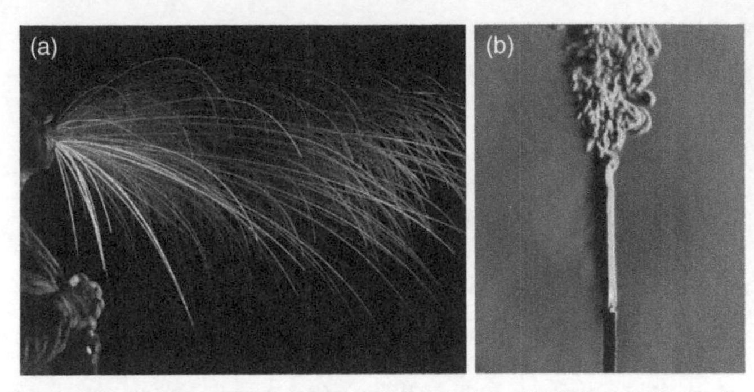

41. Two examples of topics of active research in fluid mechanics: the transmission of infectious diseases ((a) droplets ejected during a cough) and the dynamics of turbulent flows ((b) turbulent plume in air flow above a candle).

Turbulence remains indeed one the holy grails of the field, and is often quoted as being the last unsolved problem in classical physics. It is, by far, the topic within fluid mechanics where the most research is being devoted. Turbulence is everywhere in industry, from catalysis and combustion engines to flows in nuclear reactors and around aeroplanes (a turbulent plume in the air flow above a candle is shown in Figure 41(b)). But turbulence also poses the ultimate physics questions: Why do flows become turbulent? And can we predict their dynamics? Despite exciting work by generations of scientists, these questions still remain fundamentally unanswered.

The other unsolved fundamental issue in fluid mechanics concerns the mathematical structure of the equations for fluid flows (the Navier–Stokes equations, Equation (3) in Chapter 2). We do not yet fully understand if these equations are well-posed, or whether there are fundamental flaws in their mathematical structure. That problem is one of the seven Millennium Prizes Problems in mathematics proposed by the Clay Mathematics Institute at the turn of the millennium (with one of the seven, the Poincaré conjecture, already solved).

Interdisciplinary fluid mechanics

Some of the scientists with the biggest impact in the field have touched upon many different branches of fluid mechanics at once. Many were mentioned in the previous chapters. I want to close the book by highlighting the career of two modern giants of the field, different in their interests yet equipped with the same ability to make connections across fields.

Cambridge-based James Lighthill was a classically trained applied mathematician whose career ended up rivalling that of Taylor in terms of variety and impact. At some point in his life he was interested in (brace yourself): the propagation of waves, aerofoil theory, the dynamics of shocks, aerocoustics, fish locomotion, traffic flow, and meteorology. He was also unafraid of the great Reynolds number divide and did pioneering studies on the motility of cellular microorganisms in fluids. In many ways, Lighthill encompassed all of fluid mechanics.

In contrast, Paris-based Pierre-Gilles de Gennes was a theoretical physicist who was first famous for his work on superconductivity. As his career evolved, he moved closer to fluid mechanics, first thanks to his work on liquid crystals, then to polymer physics, and finally to his foray into capillary flows and wetting. Still the only Nobel Prize winner in the area of soft matter physics, de Gennes has long been the entry point to fluid mechanics for students from the world of theoretical physics who discover that much new science can be discovered in a seemingly classical field.

Future fluids

It is impossible to predict the future of any scientific field, let alone one like fluid mechanics that interacts with so many other scientific disciplines—from chemistry and biology to physics and engineering.

As a new generation of scientists embark on their careers, the 'giants' highlighted in this book have shown how fluid mechanics remains vast and varied. While these were all men, this is now beginning to shift and more women are coming into the field as undergraduates and as researchers. The future of fluid mechanics promises therefore to be more diverse and inclusive.

Just like our natural world, if we look close enough, fluid mechanics is an infinite source of new and unexpected scientific mysteries and questions.

Further reading

D. J. Acheson, *Elementary Fluid Dynamics*, Oxford University Press, 1990.

P. Ball, *Flow*, Oxford University Press, 2011.

G. K. Batchelor, *An Introduction to Fluid Dynamics*, Cambridge University Press, 1967.

G. K. Batchelor, H. K. Moffatt, and M. G. Worster, editors, *Perspectives in Fluid Dynamics: A Collective Introduction to Current Research*, Cambridge University Press, 2002.

O. Darrigol, *Worlds of Flow: A History of Hydrodynamics from the Bernoullis to Prandtl*, Oxford University Press, 2009.

P.-G. de Gennes, F. Brochard-Wyart, and D. Quéré, *Capillarity and Wetting Phenomena: Drops, Bubbles, Pearls, Waves*, Springer, 2003.

P. G. Drazin, *Introduction to Hydrodynamic Stability*, Cambridge University Press, 2002.

T. Faber, *Fluid Dynamics for Physicists*, Cambridge University Press, 2010.

E. Guyon, J. Hulin, L. Petit, and C. Mitescu, *Physical Hydrodynamics*, Oxford University Press, 2nd ed., 2015.

G. M. Homsy, editor, *Multimedia Fluid Mechanics Online*, Cambridge University Press, 2019.

J. Lighthill, *An Informal Introduction to Theoretical Fluid Mechanics*, Clarendon Press, 1986.

J. Santiago, *A First Course in Dimensional Analysis*, MIT Press, 2019.

D. J. Tritton, *Physical Fluid Dynamics*, Oxford University Press, 2nd ed., 1988.

M. Van Dyke, *An Album of Fluid Motion*, Parabolic Press, 1982.

S. Vogel, *Vital Circuits: On Pumps, Pipes, and the Workings of Circulatory Systems*, Oxford University Press, 1993.
S. Vogel, *Life in Moving Fluids*, Princeton University Press, 1996.
M. G. Worster, *Understanding Fluid Flow*, Cambridge University Press, 2010.

Index

I

K

L

M

N

P

R

S

T

V

W

SUPERCONDUCTIVITY

A Very Short Introduction

Stephen J. Blundell

Superconductivity is one of the most exciting areas of research in physics today. Outlining the history of its discovery, and the race to understand its many mysterious and counter-intuitive phenomena, this *Very Short Introduction* explains in accessible terms the theories that have been developed, and how they have influenced other areas of science, including the Higgs boson of particle physics and ideas about the early Universe. It is an engaging and informative account of a fascinating scientific detective story, and an intelligible insight into some deep and beautiful ideas of physics.

www.oup.com/vsi

RELATIVITY

A Very Short Introduction

Russell Stannard

100 years ago, Einstein's theory of relativity shattered the world of physics. Our comforting Newtonian ideas of space and time were replaced by bizarre and counterintuitive conclusions: if you move at high speed, time slows down, space squashes up and you get heavier; travel fast enough and you could weigh as much as a jumbo jet, be squashed thinner than a CD without feeling a thing - and live for ever. And that was just the Special Theory. With the General Theory came even stranger ideas of curved space-time, and changed our understanding of gravity and the cosmos. This authoritative and entertaining *Very Short Introduction* makes the theory of relativity accessible and understandable. Using very little mathematics, Russell Stannard explains the important concepts of relativity, from E=mc2 to black holes, and explores the theory's impact on science and on our understanding of the universe.

www.oup.com/vsi

Scientific Revolution

A Very Short Introduction

Lawrence M. Principe

In this *Very Short Introduction* Lawrence M. Principe explores the exciting developments in the sciences of the stars (astronomy, astrology, and cosmology), the sciences of earth (geography, geology, hydraulics, pneumatics), the sciences of matter and motion (alchemy, chemistry, kinematics, physics), the sciences of life (medicine, anatomy, biology, zoology), and much more. The story is told from the perspective of the historical characters themselves, emphasizing their background, context, reasoning, and motivations, and dispelling well-worn myths about the history of science.

www.oup.com/vsi

CHAOS

A Very Short Introduction

Leonard Smith

Our growing understanding of Chaos Theory is having fascinating applications in the real world - from technology to global warming, politics, human behaviour, and even gambling on the stock market. Leonard Smith shows that we all have an intuitive understanding of chaotic systems. He uses accessible maths and physics (replacing complex equations with simple examples like pendulums, railway lines, and tossing coins) to explain the theory, and points to numerous examples in philosophy and literature (Edgar Allen Poe, Chang-Tzu, Arthur Conan Doyle) that illuminate the problems. The beauty of fractal patterns and their relation to chaos, as well as the history of chaos, and its uses in the real world and implications for the philosophy of science are all discussed in this *Very Short Introduction*.

> '. . . Chaos . . . will give you the clearest (but not too painful idea) of the maths involved . . . There's a lot packed into this little book, and for such a technical exploration it's surprisingly readable and enjoyable - I really wanted to keep turning the pages. Smith also has some excellent words of wisdom about common misunderstandings of chaos theory . . .'
>
> **popularscience.co.uk**

www.oup.com/vsi